QA11.P585 1987

Pimm, David./Speaking mathema
Whitworth College Library

0068   00694299

SO-ATR-652

DISCARD

Speaking Mathematically

Language, Education and Society

General Editor
Michael Stubbs
Institute of Education
University of London

Speaking Mathematically

Communication in Mathematics Classrooms

David Pimm

Routledge & Kegan Paul
London and New York

First published in 1987 by
Routledge & Kegan Paul Ltd
11 New Fetter Lane, London EC4P 4EE

Published in the USA by
Routledge & Kegan Paul Inc.
in association with Methuen Inc.
29 West 35th Street, New York, NY 10001

Set in 10/12 pt Times
by Columns Ltd, Reading
and printed in Great Britain
by Billings, Worcester

© David Pimm 1987

No part of this book may be reproduced in
any form without permission from the publisher
except for the quotation of brief passages
in criticism

Library of Congress Cataloging in Publication Data
Pimm, David.
Speaking Mathematically: Communication in Mathematics Classrooms
(Language, education, and society)
Bibliography: p. –
Includes index.
1. Mathematics – Study and teaching.   2. Mathematical
notation.   3. Communication in education.   I. Title.
II. Series.
QA11.P585   1987   510'.7'1   86–24849

British Library CIP Data also available
ISBN 0-7102-1133-3

To Tam Lin's consort

Contents

Editor's preface

Simply a list of some of the questions implied by the phrase *Language, Education and Society* gives an immediate idea of the complexity, and also the fascination, of the area.

How is language related to learning? Or to intelligence? How should a teacher react to non-standard dialect in the classroom? Do regional and social accents and dialects matter? What is meant by standard English? Does it make sense to talk of 'declining standards' in language or in education? Or to talk of some children's language as 'restricted'? Do immigrant children require special language provision? How can their native languages be used as a valuable resource in schools? Can 'literacy' be equated with 'education'? Why are there so many adult illiterates in Britain and the USA? What effect has growing up with no easy access to language – for example, because a child is profoundly deaf? Why is there so much prejudice against people whose language background is odd in some way: because they are handicapped, or speak a non-standard dialect or foreign language? Why do linguistic differences lead to political violence, in Belgium, India, Wales and other parts of the world?

These are all real questions, of the kind which worry parents, teachers and policy makers, and the answer to them is complex and not at all obvious. It is such questions that authors in this series will discuss.

Language plays a central part in education. This is probably generally agreed, but there is considerable debate and confusion about the exact relationship between language and learning. Even though the importance of language is generally recognised, we still have a lot to learn about how language is related to either educational success or to intelligence and thinking. Language is also a central fact in everyone's social life. People's attitudes and most deeply held beliefs are at stake, for it is through language that personal and social identities are maintained and recognised.

People are judged, whether justly or not, by the language they speak.

Language, Education and Society is therefore an area where scholars have a responsibility to write clearly and persuasively, in order to communicate the best in recent research to as wide an audience as possible. This means not only other researchers, but also all those who are involved in educational, social and political policy-making, from individual teachers to government. It is an area where value judgments cannot be avoided. Any action that we take – or, of course, avoidance of action – has moral, social and political consequences. It is vital, therefore, that practice is informed by the best knowledge available, and that decisions affecting the futures of individual children or whole social groups are not taken merely on the basis of the all too widespread folk myths about language in society.

Linguistics, psychology and sociology are often rejected by non-specialists as jargon-ridden; or regarded as fascinating, but of no relevance to educational or social practice. But this is superficial and short-sighted: we are dealing with complex issues, which require an understanding of the general principles involved. It is bad theory to make statements about language in use which cannot be related to educational and social reality. But it is equally unsound to base beliefs and actions on anecdote, received myths and unsystematic or idiosyncratic observations.

All knowledge is value-laden: it suggests action and changes our beliefs. Change is difficult and slow, but possible nevertheless. When language in education and society is seriously and systematically studied, it becomes clear how awesomely complex is the linguistic and social knowledge of all children and adults. And with such an understanding, it becomes impossible to maintain a position of linguistic prejudice and intolerance. This may be the most important implication of a serious study of language, in our linguistically diverse modern world.

Since the late 1960s, there has been a great deal of work on classroom discourse in general and on the language of school textbooks. In this book, David Pimm takes such work a significant step further, by analysing the particular discourse, written and spoken, of the mathematics classroom. He discusses some of the fascinating connections between language and mathematics, between everyday and specialist usage, and between terminology and comprehension. There are particular relations between mathematics and written language: mathematical reasoning depends on abbreviations and symbols, which require written notation for their development and do not transfer easily to spoken language. But there are clearly also relations between talk and learning: we often grasp concepts by talking about them in our own words. Learning mathematics depends partly on learning to use such symbols and learning the meaning of specialist terms. But what is the relation between the everyday and specialist use of words? Is learning mathematics (or any other subject) just learning to talk or write like a mathematician (or other subject specialist)? Is mathematical English just a style of English? Or does it actually express different types of meanings from everyday usage? What linguistic and conceptual difficulties does learning such a variety of English involve?

Pimm's book deserves to be widely read, by mathematics teachers obviously. But for its discussion of the relations between language, logic, meaning, concepts and comprehension, it deserves to be much more widely read by educationalists, by those concerned with the psychology of learning, and by linguists.

Michael Stubbs
London

Acknowledgments

I am grateful to Ray Hemmings and Dick Tahta, editors of
Mathematics Teaching, for permission to reproduce Pimm (1984)
which appears here as Section 3 of Chapter 3, and to David
Wheeler, editor of *for the learning of mathematics*, for permission
to reproduce parts of my article 'Metaphor and analogy in
mathematics' [*flm*, 1(3), pp. 47–50] which occur in Chapters 4
and 8. I am extremely grateful to the Leapfrogs group for
allowing me to reproduce their poster of the great stellated
dodecahedron, both on the cover and on p. 35. In addition, I
wish to thank the Open University for permission to transcribe
certain videotapes in considerable detail. Finally, I am grateful to
a number of people who have shared their unpublished
transcripts with me, in particular Daphne Kerslake (pp. 59–60),
Susan Pirie (p. 120), Hiliary Shuard (pp. 65–6) and Lindsay
Taylor (p. 22 and 53).

There are many people to whom I owe a debt of gratitude as a
result of writing this book. It has taken an inordinately long time
to finish. I wish to thank publicly the following people for making
substantial comments on earlier drafts of this book; Janet Ainley,
Ann Brackenridge, Joanna Channell, Paul Ernest, Celia Hoyles,
Barbara Jaworski, Rita Nolder, Richard Noss, Susan Pirie, Neil
Ryder, Rolph Schwarzenberger and Michael Stubbs. I wish
additionally to acknowledge a debt of gratitude to John Mason
for a number of stimulating conversations about many of the
ideas contained in Chapters 2 and 3. My thanks are also due to
all my informants, both pupil and adult, in particular to Clare
Pirie, the most friendlily informative of them all. My praise must
also go to the perpetrators of the MUSE word-processing
package for the DEC-20 with which I produced the 'manu'-script.
I finally wish to record a bankruptcy-inducing debt of gratitude to
the dedicatee whose encouragement and support, both academic
and personal, allowed me to finish this book.

Preface

Prefaces, although at the beginning, are always written last.

In this book, I explore some of the language issues inherent in attempting to teach and learn mathematics in a school setting. As many of these issues arise from the fact that it is *mathematics* that is being studied, I also take a look at certain aspects of the nature of mathematics itself, not least the writing system customarily used for recording mathematics.

Pupils do not commonly hear or read much explicit mathematics outside the classroom. The teacher conventionally acts to a considerable extent as an intermediary and mediator between *pupil* and *mathematics*, in part by determining the patterns of communication in the classroom, but also by serving as a role model of a 'native speaker' of mathematics. As a consequence, one thing that pupils are learning from the teacher, then, is the range of accepted ways in which mathematics is to be communicated and discussed.

There are two major parallel strands to this book, namely examinations of spoken and written aspects of mathematics, focused on the classroom. One fundamental divide in language production is between the spoken and written channels. Many linguists have argued for the priority of spoken over written language. Stubbs (1980) disagrees, and has described the relationship between speech and writing as 'realizations of language in different media' (p. 34). In particular, these two primary embodiments of language have different strengths and demands. They also perform different functions.

While spoken language is used in a broader range of situations than is written language, there is no danger of speech having priority over writing in schools. Nor are the distinctive uses between the spoken and written channels (and the corresponding skills of listening and reading) well understood or exploited there.

Making tape-recordings (or videos) and transcripts of children talking . . . helped participants (the teachers) to realise how little *direct* knowledge they (teachers) often had, either of the features of the spoken language or of the oral abilities of their pupils. (Acland, 1984, p. 5)

The oral channel in a secondary mathematics classroom tends to be strictly controlled as a medium of communication. Part of the difficulty for mathematics involves an ethos which emphasizes a quiet, controlled, individual atmosphere as being most appropriate for its learning, not to mention its teaching.

The main theme underlying the development of this book is that of seeing mathematics as a language, and exploring the consequences of such a perception for teaching. At the outset, it may not be immediately clear how helpful this will be. One indication of this can be seen in Wheeler's provocative warning (1983a) that 'I shall keep well away from the region signposted *Mathematics is a Language*. I believe it to be uninhabited.' It will already be clear that I do not agree with this assessment, although I interpret the claim that mathematics *is* a language in a particular way, namely as a metaphor. The task I wish to perform is partially to structure the concept of *mathematics* in terms of that of *language*, but with the primary intention of illuminating mathematics teaching and learning.

The ability of this metaphor to do useful work stems from the insights provided by looking for linguistic phenomena in mathematics teaching and then applying some of the techniques of analysis of theoretical linguistics to them. One instance of this general approach can be seen in Klemme's (1981) article on speech acts in mathematics. In it, he explores some of the difficulties Dutch secondary school children had in coming to grips with the meaning of mathematical expressions, as a consequence of the abstract nature of the referents involved.

My plan is to explore particular aspects of classroom language, both of teacher and pupil, both spoken and written, and then move into a more systematic discussion of the language issues which are revealed as a result of looking at mathematics classrooms through linguistic eyes. Rather than attempt to do both at the outset, however, roughly the first half of the book concerns spoken issues and their analysis, while the latter half examines written mathematics. The topic of each chapter could easily have been worthy of a book in its own right and, on occasion, I do little more than indicate possible areas for investigation, by means of examples, questions and commentary. Wherever they appear, examples of mathematical language, whether spoken or written, are genuine, attested instances rather than items fabricated for illustrative purposes.

The following diagram attempts to summarize these two parallel, but interlocking, strands in terms of the main topics of the chapters in this book.

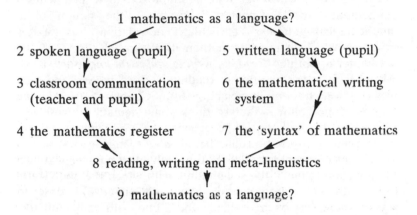

1 mathematics as a language?

2 spoken language (pupil)   5 written language (pupil)

3 classroom communication   6 the mathematical writing
  (teacher and pupil)   system

4 the mathematics register   7 the 'syntax' of mathematics

8 reading, writing and meta-linguistics

9 mathematics as a language?

Following an introductory chapter which provides some examples of the range of issues involved, Chapter 2 addresses the topic of pupils' mathematical talk in its various incarnations. This includes a short account of the notion of *discussion* in mathematics, a topic which is attracting much attention currently. Chapter 3 examines some of the common styles of spoken interaction to be found in many mathematics classrooms, by means of transcripts and their analysis. The place of mathematics in relation to a natural language such as English is the theme of Chapter 4 which introduces both the technical linguistic term *register* to help describe the relationship, as well as the concept of *metaphor* in mathematics.

My involvement with metaphor pervades this work, reflecting a belief that it is this phenomenon most of all which enables natural language to function as the powerfully expressive medium it is. It is by no means widely accepted, however, that metaphor has anything to do with mathematics. One initial inclination might be to view mathematics and, hence, mathematical language, as clear and precise, while metaphors seem fuzzy and, at best, suggestive. However, I argue in Chapters 4 and 8 that the process of metaphor is as important and ubiquitous in mathematics as it is in natural language.

In Chapter 5, I turn my attention to written aspects of mathematics; in particular, to the range of written forms in which pupils express mathematical ideas and especially to the development of algebraic symbolism which is so often taken as the hallmark of written mathematical language. Chapter 6 explores the nature of the symbol systems employed in formal written mathematics, while Chapter 7 examines the notion of a 'mathematical grammar', particularly in relation to computer programs which manipulate mathematical symbols.

Chapter 8, entitled *Reading, writing and meta-linguistics*, takes a brief look at the question of reading in mathematics, and then picks up the notion of metaphor from Chapter 4 to examine aspects of metaphor in the notations of mathematics themselves. I conclude by looking at the growing area of meta-linguistics (concerned with intuitions and beliefs *about* language itself) and its relevance to mathematics. In the final, short chapter, I return to an examination of the same theme with which the book starts (namely the perception of mathematics as a language) in order to assess where has been reached, and I end with some remarks about the nature of mathematics education itself.

It would be remiss to finish any introduction to a topic concerned with mathematics education without reference to the Cockcroft report *Mathematics Counts* (HMSO, 1982). The frequency of its citation has a similarity to director Alfred Hitchcock's appearance in all of his films. Hitchcock became concerned in his later career that his audience was more involved initially in watching for his increasingly fleeting appearances than with the developing narrative (Truffaut, 1967). Consequently, he resolved to appear as near the beginning of the film as was possible in order to minimize the distraction. Similarly, once Cockcroft has been quoted, the reader's attention can be returned to its proper place:

. . . mathematics provides a means of communication which is powerful, concise and unambiguous. Even though many of those who consider mathematics to be useful would probably not express the reason in these terms, we believe that it is the fact that mathematics can be used as a powerful means of communication which provides the principal reason for teaching mathematics to all children. (para. 3)

The report, unfortunately, did not then discuss in detail *what* it was that was to be communicated, but mathematics can be profitably seen as both medium and message, with the two often inextricably and deliberately mixed. While I take exception to the word *unambiguous* in the above, this quotation serves the purpose of introducing *communication* as one of the central concerns of anyone interested in mathematics education. Mathematics is, among other things, a social activity, deeply concerned with communication. In this way, 'We find an echo of K. C. Hammer's aphorism that the most neglected existence theorem in mathematics is the existence of people' (Brookes, 1970, p. vii).

David Pimm
Milton Keynes, May 1986

1

Mathematics as a language?

Sellers: How's your Mathematics?
Secombe: I speak it like a native.

                  Spike Milligan, *The Goon Show Scripts*

Without aspiring to be one of those people who analyses or explains jokes, I nonetheless want to begin by posing the question 'why does this joke work?' – that is, why does it work better with *mathematics* than, say, with *geography* or *woodwork*. The expression 'I speak it like a native' is used only of a language which is *not* one's native language. Yet its use suggests the existence of native speakers with whom a comparison can be made, and the participants in the exchange are clearly construing the term *Mathematics* as the name of a language.

What is incongruous about the idea of a native speaker of mathematics? Included among various possibilities are the following. Firstly, perhaps it seems odd because mathematics is commonly viewed as the epitome of something which, if successfully learnt at all, is learnt overtly and consciously. This is in marked contrast to the seemingly effortless success with which virtually every child acquires language. Secondly, perhaps it is because mathematics is perceived as overwhelmingly *written*, involving unfamiliar symbols, rather than words: something that is to be done on paper, rather than a means of oral communication.

On the first point, it has been ironically suggested that if children had to be taught to speak using the methods by which the majority are taught to read and write, many might not say anything. Although it could be claimed that acquiring a first language is unlike learning anything else, both Gattegno (1970a) and Papert (1980) have written extensively on the need to model learning in general, and mathematics learning in particular, on human strengths such as the ability to acquire a first language.

1

In particular, in his book *Mindstorms*, Papert describes the computer language LOGO as a means for children to gain entry into environments (which he calls 'Mathland') where mathematics is the everyday language of discourse. It is by means of communication with the machine, initially through control of a screen cursor known animistically as *a turtle*, that 'speaking' mathematics becomes a natural activity. It is one peculiar consequence of the sense of immediacy and involvement which can be generated by a computer presence that, despite its totally written characteristics (e.g. the keyboard or screen text), interacting with such a machine is customarily referred to as 'talking to it' (or, perhaps, 'telling it what to do', given the imperative terminology of *instructions* or *commands*).

In response to the second point, which invoked a view of mathematics as predominantly written, many languages exist which have no written form; but is mathematics really an example of the converse? Is the writing system used for mathematics its most dominant feature with no spoken analogue? Surely not, for when written mathematics is read aloud, even work presented completely in symbols, it can emerge as spoken English, or any other natural language with a sufficiently developed mathematical register. Thus written symbolic mathematics can be projected into many natural languages.

In support of Secombe, it should be pointed out that learning mathematics can well be likened to learning a *foreign* language. Certainly Goethe thought so. In a collection of his writings entitled *Principalen und Maximen*, he complained that, 'Mathematicians are a sort of Frenchman. Whenever you say anything to them, they translate it into their own language and right away it is something completely different.' For many people, mathematics is incomprehensible in a way akin to a foreign language which they do not speak.

My aim in this book is to take seriously the often-heard claim that 'mathematics is a language', and to explore what some of the consequences of so doing are. What sense can be made of this claim? The use of *is* could be taken to indicate a definition, comparable with the statement, *a function is a set of ordered pairs*. Equally, it could be interpreted in a descriptive, predicate sense, such as *this chair is a comfortable one*. It is also possible, however, to construe such an expression as a metaphor, more akin to *George is a lion*, and this is my intention here. This last expression could be used to mean many things, other than its simple literal meaning, concerning a lion named George. Possibilities, apart from the familiar one that he is brave, include: George sleeps about twenty-one hours a day; he is a carnivore; he has a tremendous head of (ginger) hair (a *mane*); or that he roars a lot ('be careful how you approach him').

There are, equally, many possible interpretations of the above claim about mathematics and in this chapter I start to describe some of them. It is important to be aware that metaphors are inherently conjectural. They do not have to be accepted as useful and there are several intellectual strategies available which provide defences against them. One way of examining the potential of such a metaphoric claim is to explore its consequences.

I am aware that employing this perception (that of seeing mathematics, and hence its teaching and learning, in linguistic terms) as a controlling metaphor will distort. All metaphors do, by stressing certain things at the expense of others; but this is one means by which they function successfully. In ascertaining how applicable this particular metaphor is, I hope to illuminate certain aspects of mathematics and, more important, its teaching, which may not have previously been apparent. Before starting to unravel the nature of the metaphor, however, I wish to examine briefly what is involved in knowing any language.

1 What does it mean to know a language?

There are many things which native speakers of a language are able to do which, for example, allow language to be discerned in the noise of speech. At one level, as Whitehead (1969, p. 308) claimed, 'spoken language is merely a series of squeaks'. The ability of human beings to make sense of these noises, however, indicates that much more is involved.

The purpose of this section is to try to summarize very briefly certain aspects of language structure and function. Among the more obvious general attributes which enable us to use language fluently are aural comprehension and speech on the one hand, and reading and writing on the other. These larger-scale abilities in turn include, among others, a knowledge of spelling, pronunciation, syntax, and the possession of a vocabulary, together with a detailed knowledge of its structure. As a more

subtle instance, part of knowing a language is precisely the ability
to segment a continuous outpouring of sound into individual
words. To refer to this passively as 'hearing' the words seems less
appropriate than describing it as successfully *imposing* a structure
of words *onto* a flood of sound. Listening to an unfamiliar
language highlights this, in that the hearer does not even know
which features to attend to for cueing information about how to
divide up the apparently continuous flow of sound.

As a second example of a subtle linguistic ability, linguists refer
to the idea of *communicative competence*. The general notion of
communicative competence involves knowing how to use language
to communicate in various social situations – how to use language
appropriate to context. In other words, it requires an awareness
of the particular, conversational or written, context-dependent
conventions operating, how they influence what is being communicated, and how to employ them appropriately according to
context. Stubbs (1980, p. 115) writes: 'A general principle in
teaching any kind of communicative competence, spoken or
written, is that the speaking, listening, writing or reading should
have some genuine communicative purpose.' In passing, let me
raise the question of whether this is the case with mathematics
teaching, particularly when teachers ask pupils to record. Is the
teacher clear about the purpose of making written records and is
this intention clear to the pupils?

Communicative competence, then, involves knowing how to
use and comprehend styles of language appropriate to particular
social circumstances. Searle (1979) uses the expression 'It's
getting hot in here', to illustrate a wide range of possible
meanings variously cued according to context, intonation and
implication. For instance, this expression could be employed
metaphorically as a comment on the progress of a meeting, or as
an oblique request to open a window.

Such indirect commands by teachers are commonplace in
school settings. The surface form of the comment may not be
marked as a question or a command, but that is how it is
intended. The apparent inquiry, 'Now which table is ready to go
out?', is actually intended as the command, 'hurry up, tidy up
and be ready to go out.' Other attested classroom instances
include 'I hear talking', as an oblique instruction to be quiet, and
'We still haven't tidied up our toys', as an indirect command to
do so.

Barnes (1976, p. 16) provides an instance where a teacher interrupted a pupil who had just started to read aloud with the remark 'Begin again, John. No-one can hear you', where he actually meant 'speak up'. As Barnes points out, at the literal level of meaning, the teacher's comment was clearly false. Stubbs (1983a, p. 40) discusses the apparent request 'Right, fags out please' (said by an English teacher to a class in which no-one was smoking) in terms of a signal opening the channel of communication and attempting to gain the pupils' attention.

While not suggesting that the conventions in mathematics classrooms are necessarily different from those in school generally, I want to ask whether there are any specifically mathematical instances of this sort. To what extent is a mathematics classroom perceived either by teachers or pupils as a context containing specific linguistic conventions which govern the sending and receiving of messages? One example of this, related to the mathematics classroom use of the pronoun *we*, is discussed in Chapter 3.

The most crucial linguistic abilities, however, are those of being able to assign meaning to what is heard or read, and of conveying one's intentions through the spoken and written channels. As the linguist Jakobson put it, 'Language without meaning is meaningless.' Yet there are some deep difficulties here, highlighted by the ability of computers to mimic human communication. Certain dialects of LOGO apparently treat procedures as verbs, e.g. to SQUARE or to SPIRAL (thereby enabling procedure calls by name to have the form of imperatives), and if an undefined procedure is requested, error messages such as 'I don't know how to SQUARE' appear on the screen, which suggest communication with an intelligent, English-speaking entity. Typically, the turtle is anthropomorphized for children by adults using the metaphor of 'you must use commands the turtle understands', and if not, 'you have to teach the turtle the meaning of those words'. Hence, the language of 'the turtle doesn't know how to . . .'.

A second problematic example comes about from language-manipulating computer programs such as the pseudo-psychoanalytic one called ELIZA (Weizenbaum, 1984). In these instances, I feel that ascription of concepts such as *meaning* or *understanding* to the computer is highly inappropriate. Because *we* can assign meanings and interpretations to the marks appearing on the screen, the presumption seems to be that the agency producing them also had a similar comprehension and intention.

Despite this difficulty, sending and receiving comprehensible messages appropriate to the context is the primary point of language, a point which highlights an inquiry about mathematics. My first question concerns whether or not these are recognized as appropriate aims for a mathematics teacher. Easley and Easley (1982) report their experience in an unusual Japanese primary school, one of whose declared *mathematical* aims was the oral fluency of their pupils, 'to help children learn to speak clearly and with confidence'. The Easleys add, as an aside, 'which didn't seem like a normal goal for mathematics classes'. If we are to view mathematics as a language, communicative competence becomes an important consideration, and meaningful communication an overwhelming concern.

Being fluent in a language, then, involves the ability to tap into the resources implicit in it and to use these potentialities for one's own ends. Halliday (1978), for example, writes of knowing a language in terms of access to and mastery of three interlocking systems, namely the *forms*, the *meanings* and the *functions*. The last category involves the use of language to do specific things, for instance, the various ways in which it is possible to act on the world 'at a distance'. When a baby wishes to operate on the world to achieve its various ends, its first contact is direct and physical. The development of spoken language permits certain ends to be achieved indirectly, e.g. by asking for or demanding them. Further control of written language again broadens the range of possibilities which are now 'within reach'. None of us ever stops learning a language to some extent, as our sense of control and mastery of these systems deepens in the light of wider experience.

Computer languages can be seen in this light. Knowledge of a computer language affords control over various 'screen objects' for instance, but this control is also very much one of action at a distance. No direct physical manipulation is possible, the only access is via the computer keyboard. The desire to be able to interact with these screen objects (I am particularly thinking of LOGO's turtles and sprites) provides a strong and genuine motivation for struggling with the syntactic complexities of a formal computer language. Papert refers to these screen phenomena as 'transitional objects'. For me, they are transitional in two senses: firstly, on a continuum from concrete to abstract, and secondly on a continuum from public to private.

Screen objects such as the turtle are fairly abstract, in that they require a formal language to manipulate them; yet they are still publicly accessible, unlike an individual's thoughts. The last function of language I wish to mention here is that of using language to gain a greater access to and control over one's thoughts. For Papert, this is the central contribution of learning to program a microcomputer. These points will be explored at greater length in subsequent chapters.

2 Pursuing the metaphor 'mathematics is a language'

How might such a systematic exploration of the metaphor *mathematics is a language* be undertaken? Just as with the lion example, one starting point might be to list some of the main components or attributes of a natural language such as English, as well as some of the functions to which humans put it, and start looking for mathematical analogues in school settings. In what follows, I attempt to exemplify some of the linguistic phenomena which can be found within the context of mathematics classrooms, in order to provide some feeling for the possible range involved.

The remainder of this chapter is organized under the three general headings of *meaning, symbols and the things symbolized* and *syntax*. The examples provided are intended as a foretaste of some of the topics to be treated as the book develops, as well as an illustration of some of the potential scope for exploration. I hope that it will become clear that I have only scratched the surface of possible areas for inquiry.

Meaning

According to Thom (1973), a prominent contemporary mathematician, the construction of meaning rather than the question of rigour is the central problem facing mathematics education. How is such meaning to be constructed? A similar question can be asked about young children learning to convey and derive meaning by means of spoken or written language, including the meaning of the component words as well as the information coded structurally (for instance, in the order of the words). Children, in common with the rest of us, try to make sense of whatever they hear or read. The search for meaning can lead to some unusual conclusions. Words are to be found in dictionaries, suggesting that they provide an appropriate unit for the discussion of meaning. It is commonplace to hear a teacher in a classroom asking pupils if they have understood the meaning of a particular word, and possibly trying to test their understanding of it by requesting either a formal definition or a paraphrase of its meaning. Mathematical discourse is notorious for involving both specialized terms and different meanings attached to everyday words.

Are there some confusions in mathematics classes at the level of interpretation of particular words? Below are a couple of instances where the mathematical usage of a certain term has resulted in an alteration in meaning or connotation of an ordinary English word. If the hearer is unaware of this variant usage, resulting in the everyday meaning being carried over to the mathematical setting, a number of understandable difficulties may ensue.

In response to the written question, 'What is the difference between 24 and 9?', one nine-year-old replied, 'One's even and the other's odd', whereas another said, 'One has two numbers in it and the other has one.' These responses suggest a failure to comprehend the term *difference* as being used in a mathematical sense whose meaning involves the notion of subtraction. There is also a second distinction between the two responses (to do with the perceived level of the difference) which will be discussed shortly.

If something is a *fraction* of its former cost, it is certainly not expected to be more expensive, yet not all mathematical fractions lie between zero and one. Consider the injunction to 'go forth and multiply' – the very notion of multiplication seems to involve making something bigger. In an interview with a twelve-year-old pupil concerning the multiplication of decimals, Swan (1982) recorded the following exchange. (Pupil comments are italicized.)

6.23 × 0.48
What can you tell me about the answer to that? What sort of answer do you think it will be?
About eight. Yes, it's not going to be a lot more than that. It's

going to be slightly larger than the 6.23, but seeing how it is not even a whole number, it can't work out very large, say about eight.

Below is a second interview carried out by Swan (1981) with a fourteen-year-old boy, concerning the same problem.

> How about if I did one like this?
>   6.23 × 0.48
> What do you think the answer to this would be?
> *Um. About twelve. Yes.*
> How did you get that?
> *It's about half of a whole number. Halfs into six equals twelve.*
> In this case the answer is twelve. You've divided, but the answer is bigger. What if you multiply? Can you make the answer smaller?
> *No, it won't work out smaller.*
> Do you want to work it out? (Hands him a calculator.)
> *(Laughs).*
> What's happened?
> *It's got smaller. It's 2.2904.*
> How do you explain that then?
> *I don't know. I thought maybe if you . . . Oh, is it one of those funny numbers? But multiplication still makes it bigger.*

At an elementary level, 'multiplication makes bigger' expresses a valid generalization about the operation of multiplication when applied to whole numbers. When the notion is extended, and the same words and symbols are applied metaphorically to the new situation (either to fractions or to negative numbers, for example), this observation makes sense, but it is no longer true. The failure of a valid generalization, one which accords with the everyday connotations of the word *multiply*, can result from not perceiving the novel use to which the words are being put.

The concept of *number* itself undergoes a considerable expansion in meaning as pupils proceed through the school system. At certain stages, pupils will probably have heard the statement that, 'all numbers are either odd or even'. On other occasions, they are likely to have been informed that fractions are numbers. Combining these two pieces of information requires sense to be made of a fraction being odd or even. The following extract (Harvey, 1983, p. 28) illustrates two pupils, David (D) and Robert (R), successfully finding such a context. RH is the questioner.

D: Fifteen's odd and a half's even.
RH: Fifteen's odd and a half's even? Is it?
D: Yes.
RH: Why is a half even?
D: Because, erm, a quarter's odd and a half must be even.
RH: Why is a quarter odd?
D: Because it's only three.
RH: What's only three?
D: A quarter.
RH: A quarter's only three?
D: That's what I did in my division.
R: Yes, there's three parts in a quarter like on a clock. It goes five, ten, fifteen.
RH: Oh, I see.
R: There's only three parts in it.

The above extract illustrates one means by which pupils construct mathematical meaning. A context is looked for and found in which the existing definitions of *odd* and *even* can be sensibly interpreted. There seems to be a considerable weave of prior mathematical knowledge being employed to provide consistency. One instance is David's possible use of the fact that any two odd numbers, when added together, always produce an even number, when he claims, 'a quarter's odd and a half *must* be even' (my emphasis). It is not clear how a teacher might proceed to convince or persuade David to accept the more conventional opinion of the inappropriateness of the extension of the notions of *oddness* and *evenness* to fractions.

Apart from differences in everyday usage of individual words, far more important for the construction of meaning in mathematics seems to be the analogue of the process of metaphor. One initial reaction to metaphors in mathematics may be to consider them unreliable, and hence unsuited to mathematical work. I think there are important cases of metaphor to be found in algebra and even in arithmetic, and I wish to claim that metaphor is as central to the expression of mathematical meaning, as it is to the expression of meaning in everyday language.

Is metaphor a very sophisticated notion, poorly understood and little employed by even teenage pupils? On the contrary, very young children are capable of its use. Winner (1979) explores the possibility of interpreting young children's over-extension of the application of certain terms as an early use of metaphor. She cites what seems to be a clear example of an action described metaphorically, namely three-year-old Adam commenting on his replacing the paper cover on a crayon after it had come off, by saying, 'I putting on your clothes, crayon.' If you do not know the appropriate word for something (e.g. *cover*, here), metaphor is one way of coping with the situation, and the choice of substitute term is determined by how you see the situation.

Here is a second example, indicating that it is metaphor and not just simile which is intended.

> A five-year-old boy returning from a birthday party announced with wide eyes and forceful voice: 'and the cake was a castle!' When offered an alternative version, that the cake was like a castle, he replied emphatically, 'No, it was a castle, but it was built of other things.' (Mason *et al.*, 1985)

As a further instance of metaphoric imagery of description employed by a young child, this time in a mathematical context, Tahta (1970, p. 27) reports the comments of a seven-year-old girl with whom he was talking (her remarks are italicized).

> I had shown [her] a circle with some lines, red ones intersecting the circle and green ones not doing so.
> What could you say about the red lines?
> *Well they are fighting – sort of cutting up the circle.*
> And the green lines?
> *They are protecting – yes, they are guarding.*
> I had then drawn a tangent in pencil and asked what colour it might be given. There was a pause. Then:
> *It's green escaping from red.*
> A brief pause.
> *Or red escaping from green.*
> Another pause.
> *Oh dear, it's a helpless man.*

Who is to say without examining the idea further that perceiving geometric incidence and proximity in terms of threat and attack is not a *mathematically* useful way of viewing such situations? As it happens, this very notion has recently been explored on film and in the classroom by Lemay (1983). Among the other things which younger children have yet to learn is the manner in which it is deemed seemly to discuss mathematics. Widespread use of emotionally-tinged metaphor (whether or not cued in by the use of colours) is not commonly accepted as one of them.

There are styles of communication which are *conventionally* considered appropriate to particular disciplines. One written reflection of this is the stylized and constricted nature of verbal problems which often form the most difficult arithmetic tasks, from the pupils' perspective, that they are called upon to perform. Nesher (1984) has claimed that pupils are fully aware of this restricted form or template into which such problems are customarily squeezed. One means of exploration she employed was to ask pupils to propose such problems of their own which would result in a particular 'sum', such as $6 + 2 = 8$. Scant attention was paid to the plausibility of the proposed context. One suggested example was 'My mother had six irons. She was given another two. How many irons did she have altogether?'. All that seemed to be important was the *form* of the question.

Working in the context of decimal problems, Swan (1982) found even more semantically-anomalous instances of proposed contexts. Below are some of the written responses to the following task that he received from a class of twelve-year-olds.

Write your own story to go with this sum
    $4.6 + 5.3 = 9.9$

Tony had 4 portions of cake plus a 6th of a piece. His mother gave him 5 portions and a third of a portion. Tony had eaten 9 portions and a 9th of a portion.

James had 4.6 sweets. His best friend gave him 5.3 sweets and he has 9.9 sweets altogether.

Dad gave me £4.6p. Mom gave me £5.3p. I had altogether £9.9p.

John had 4.6 video tapes he sold them and had enough money

to buy 5.3 bags of sweets and he then calculated up how much he had and he had 9.9.

John had 4.6 pages of a book left to read and his father had 5.3 pages to read so between them they had 9.9 pages left to read.

One, in a more modern idiom, ran:

John had a computer. His father made him a 4.6 bit rampack but John wanted more so his father made him a further 5.3 bit rampack which made 9.9 bits when he coupled it to the 4.6 bits of ram he already had.

Aside from the interesting instances of interpretation of what are and what are not decimal numbers (money, for instance, or the fractional conversion implicit in the first proposed pupil problem), these stories all exhibit the apparent irrelevance at one level of the surrounding story in mathematics classes. The stories do not have to be plausible or even make sense provided they contain the requisite numbers and a guide to the operation.

The possibility of being diverted from the mathematical structure of the problem and engaging with the narrative content of these situations has been beautifully captured by author Philip Roth (1970), describing the problems his father used to pose him as a child.

'Marking Down', he would say, not unlike a recitation student announcing the title of a poem. 'A clothing dealer, trying to dispose of an overcoat cut in last year's style, marked it down from the original price of thirty dollars to twenty-four. Failing to make a sale, he reduced the price to nineteen dollars and twenty cents. Again he found no takers, so he tried another price reduction and this time sold it . . . All right, Nathan, what was the selling price if the last markdown was consistent with the others?' Or, 'Making a chain.' 'A lumberjack has six sections of chain, each consisting of four links. If the cost of cutting open a link . . .' and so on.

The next day, while my mother whistled Gershwin and laundered my father's shirts, I would daydream in my bed about the clothing dealer and the lumberjack. To whom had the haberdasher finally sold the overcoat? Did the man who bought it realize it was cut in last year's style? If he wore it to a restaurant, would people laugh? And what did last year's style look like anyway? 'Again he found no takers', I would say aloud, finding much to feel melancholy about in that idea. I still remember how charged for me was that word *takers*. Could it have been the lumberjack with his six sections of chain who, in his rustic innocence, had bought the overcoat cut in last year's style? And why suddenly did he need an overcoat? Invited to a fancy ball? By whom? . . .

My father . . . was disheartened to find me intrigued by fantasies and irrelevant details of geography and personality and intention, instead of the simple beauty of the arithmetic solution. He did not think that was intelligent of me, and he was right.

This ability to throw away the outer container of 'irrelevant' information in an essentially arithmetic task is a skill which teachers attempt to teach in schools. Yet the contexts in which these problems are embedded do carry considerable social implications for the perceived uses and usefulness of mathematics. Maxwell (1985), in an article entitled 'Hidden messages', juxtaposes examples of these material-world containers for arithmetic problems from different countries and cultures in an attempt to draw attention to some of the political implications and social values implicit in them.

The notion of meaning is a very complex one, and contains many more problems than might have been thought at first sight. By means of the foregoing collection of instances, I have tried to indicate some of the possible sources of difficulty which might arise. In later chapters, particularly Chapter 4, they will be examined in more detail. Not all difficulties, however, have to do directly with confusions over meanings. In the next section, I outline some problems which arise as a consequence of the abstract nature of mathematical 'objects' and the confusions between the signs and symbols of mathematical language, on the one hand, and these objects themselves, on the other.

Symbols and the things symbolized

In normal written and spoken communication, the words or sounds employed are playing a secondary role, often referred to as being the 'carrier' of the message. The symbols in ordinary language, namely sounds and written marks, are certainly not the customary focus of attention. In order to be able to talk *about* the language being employed, by means of that language itself, certain devices must be employed in order to distinguish between, for example, a word being used in order to communicate, and the same word being employed as the *object* of the communication.

One such written device is the use of italics. Thus, the sentences

Mathematics should have a place in every child's education
and
Mathematics has eleven letters

employ the word *mathematics* in very different ways. The second sentence cites the word as an object, rather than being linked to its referent in the customary manner. These distinct uses of words reflect the difference between discussing attributes of the 'container' of the idea and those of the idea itself. This way of speaking about the relationship between symbols and their meanings reflects the *conduit metaphor* (Reddy, 1979), which identifies linguistic expressions as containers for ideas, and communication as their transmission.

The ability to construct sentences in English about expressions in English is the source of many jokes, which function on the basis of the listener being able to switch levels and construe the symbolic form as an object with properties in its own right.

Q: What is the longest word in the English language?
A: *Smiles*, because there is a mile between its first and last letters.

Such potential confusion of levels arises regularly in mathematics, though it is seldom exploited for humorous purposes. On the contrary, because it is less apparent, it is far less easily and immediately recognizable. Mathematics seems peculiarly prone to confusions between the symbols through which ideas are communicated and the ideas themselves. In part, these difficulties arise from a confusion of levels, exacerbated by the attempt of mathematicians to reflect relationships among the ideas by means of relationships among the symbols. For instance, the value of a digit in a place-value numeration system such as ours is reflected in its left-right position on the page in relation to the other digits present. A second example employing the same surface distinction is the representation of the magnitude of a whole number in terms of its left-to-right placing on a horizontal number line. In the latter case, however, if *a* occurs to the right of *b* on the number line, then *a* is larger than *b*.

There are no direct links between the symbol and the meaning of the Hindu-Arabic numerals (the familiar symbols 0–9) in use in the West today. The link between number and numeral in this system is *conventional*. This is not the case with some earlier systems which used a principle of repetition of symbols in a non-place-value manner. Thus, for instance, the Egyptian hieroglyphic numerals reflect their meaning in the sense that there are six *ten* symbols in the symbol for sixty. This transparency with regard to meaning is one reason for the suggestion that Egyptian hieroglyphic numerals would make a useful intermediate numeration system for young children to learn before coming to grips with the greater abstractness of our full-blown decimal place-value system.

The existence of such form–meaning links in certain circumstances, however, might also suggest that this is the expected relationship between mathematical symbols and their meanings. It could therefore result in attempts to look for iconic links between symbol and object. In a *Peanuts* cartoon by Schulz, one of the characters in a classroom claims:

> Anything with a '3' is easy because you just take the first number and then count the little pointy things on the '3' and you have the answer.

She then goes on to demonstrate this rule.

> Let's see . . . '9 + 3' . . . I take the nine and count the little pointy things on the three . . . ten, eleven, twelve . . . the answer is 'twelve' . . . Ha!!

There are a number of confusions about the use of numbers in everyday life. With ages, it is frequently unclear to young children *what* is being counted. It is quite common to hear a child claim that he or she is five, and no one else is allowed to be five, because *they* are. The identification is complete. House numbers have no interpretation in a cardinal (counting) sense. Even the ordinal link in a street numbered in the familiar fashion of *odd* on one side of the street and *even* on the other, is not immediately apparent. In such circumstances, it is again common to identify the number with the object it is labelling. The referent for 15 *is* the particular house itself. A five-year-old, rugby-mad, was being asked if he knew his numbers. The interrogator pointed to a 9 and asked 'What's that?' (presumably hoping for an answer like, 'That's a nine'). The answer immediately came, 'Gareth Edwards'.

In a short extract entitled 'What can you tell me about these?', Skett (1985) records the responses of a six-year-old, Michael, when shown the numerals 7, 4, 3. The variety of response reflects him attending to them as mere marks, as symbols (with names) and as meanings.

> Some of them are the wrong way round – no they're not, but the 3 is the odd one out because the others have got something straight on them.
> 7 and 3 is seventy-three – put it like this, 3 and 7, and it is thirty-seven.
> They all begin with a different letter.
> It could be 4,000 if you had three noughts.
> And, ah I got it, 4 + 3 = 7.

The fact that certain descriptive adjectives relating to the symbolic form are commonly used apparently describing attributes of the referent itself contributes to this confusion between symbol and object. Are there different types of numbers called *binary*, *decimal* and *fractional*, in the same way that there are *prime*, *whole* and *triangular* numbers? Or are the former categorizations based on surface phenomena, namely certain aspects of the *forms* of the conventional symbols rather than involving a property of the number *per se*? In the earlier cited example, concerning the mathematical use of the word *difference*, one pupil had observed a distinction between the numbers themselves (oddness and evenness), while the other had found a distinction between the formation of the numerals representing the numbers. Algorithms are frequently taught in terms of operations to be caried out on the surface form, with injunctions such as *invert and multiply* or *to multiply by ten add a nought*. In this way, pupils are encouraged to see the symbols as the objects under consideration in mathematics classes. Is 7 bigger than 2? Take out your ruler and see!

Here are a couple of exchanges between teachers and pupils which reflect this confusion. In response to the question 'how many fours in twenty-four?', a ten-year-old pupil answered 'one'. The word *in* was apparently interpreted as meaning 'making up' (as with the number of letters in a word), rather than 'going into' (an expression used to indicate division). The second interchange reflects a symbolic confusion, to do with the common use of letters as mathematical symbols:

Teacher: Let *n* be a number.
Pupil: But *n* is a letter!

Algebra is frequently, but erroneously, thought of in terms of calculations with letters, where the letters are *letters* rather than convenient symbols which we know how to form and distinguish from one another. A mathematical 'joke' similar to the one about the longest English word runs as follows:

Q: What is the value of the product
     $(x - a)(x - b) \ldots (x - z)$?
A: Zero.

The mathematical expectations engendered by such expressions is that the answer will be found by expansion of the brackets, followed by subsequent factorization into some simplified form. Certainly students to whom I have presented this as a question immediately start writing down a polynomial expression involving x^{26} — the sum of all the letters of the alphabet times $x^{25} + \ldots$. On offering an answer of zero (consider the twenty-fourth bracket of the product), a confusion with regard to the role and function of symbols can be observed. One of the many conventions regarding the choice of symbols for variables is that a variable name must not be the same as any member of its replacement set. Yet this is a common phenomenon both in English and mathematics (see Mason and Pimm, 1984).

In summary, by means of this collection of examples gathered under the heading of *symbols and the things symbolized*, I have attempted to highlight some of the confusions which can occur when the attention of the pupil is focused more closely on the symbols themselves (i.e. the language itself), rather than on the meanings of those symbols. This problem arises partly because of the abstractness of the referents for many of these symbols. It also comes about from an attempt to teach to pupils the practice of successful mathematicians, who act upon the symbols *as if* they were the mathematical objects themselves. This practice permits very successful and fluent computations to be made, but is done at the cost of enhancing the risk referred to above. It is this last point, the possibility of operating purely syntactically on mathematical symbols, that I turn to for my third collection of linguistic phenomena in mathematics.

Syntax

Syntactic rules reflect grammatical relationships among words, phrases and sentences. It is possible to formulate transformations which account for the perceived meaning relationship between many English sentences; for example, active and passive, dative movement or question formation. Here are some elementary examples of sentence pairs, similar in meaning, related by the above transformations.

Mike cooked the dinner.	The dinner was cooked by Mike.
I gave the book to Joanna.	I gave Joanna the book.
The book is here.	Is the book here?

It is equally feasible to formulate some of the transformations in mathematics in an analogous fashion, in which case algebra can be seen as symbol manipulation according to given rules, the grammar of symbolic expressions.

$$a \times b = b \times a$$
$$(a + b)^2 = a^2 + 2ab + b^2$$
$$a(b + c) = ab + ac$$

While much second language teaching has moved away from the explicit teaching of rules for the construction of various forms (such as the passive or the negative), this is not the case with mathematics. Many of the algebraic algorithms are verbally coded in terms of concise precepts dealing with the surface form, that is, operating at the syntactic level of symbols only. Such precepts, which are consciously and deliberately taught, include:

collect all the *x*s on one side;
take it over to the other side and change the sign;
do the same thing to both sides.

Also, certain perceived errors in mathematics can be accounted for in terms of the operation of over-generalized transformations, just as the syntactic formulation of children's variants of adult forms has done. For instance, the commonly-heard variant form of 'I goed' can be accounted for in terms of an over-generalization of the rule which adds an '-ed' to the verb to form the past tense. The common, erroneous expansions of $(a + b)^2$ as $a^2 + b^2$ and $\sqrt{a + b}$ as $\sqrt{a} + \sqrt{b}$ can be similarly viewed in terms of an over-extended distributivity principle.

However, it seems to me that many of the errors that occur in algebra do so precisely because algebra is frequently approached in such an abstract symbol-manipulating manner, without any regard for possible meaning. Much of the computational fluency which mathematicians achieve with symbolic manipulation arises as a result of being able to work solely on the symbols themselves without thinking about their meanings. The crucially important skill, however, is the ability to reintegrate the symbols with their meanings *at will*, in order to interpret or check the details or results of such symbolic calculations.

Summary

There has long been a concern with linguistic aspects of mathematics and the influence they might have on its teaching. (For a general review of work in the area of language and mathematics, see Austin and Howson, 1979.) In this opening chapter, I have begun to explore the apparent claim that *mathematics is a language*, by construing it as a metaphor. Having briefly expanded the concept of knowing a language in order to identify some of its component abilities, I concluded by examining a range of difficulties which can occur in the mathematics classroom under the three headings of *meaning, symbols and the things symbolized* and *syntax.*

Subsequent chapters will take up one or more of the aspects raised in this opening discussion, and explore them in detail, with language structure, meaning and use all being of interest. The next chapter concerns the issue of spoken mathematics, about which little has been explicitly said so far. There are important differences between oral and written language which require examination in relation to how they impinge on mathematics. The current call for more discussion in mathematics classes draws attention to the need to redress the perceived imbalance between these two modes of communication.

2

Pupils' mathematical talk

Communication is not the only function of language.
Douglas Barnes, *From Communication to Curriculum*

'Who needs the most practice talking in school? Who gets the most?' Exactly. The children need it, the teacher gets it.
John Holt, *How Children Learn*

Like much informal talk, spontaneous discourse about mathematics is full of half-finished and vague utterances. The following excerpt from a secondary class working on the area of similar figures (rectangles, under various scalings of the sides) provides an example. The teacher (T) is drawing on the board during the first few remarks; P is one particular pupil.

T: Let's see, here's my – here's this – all the lengths reduced by three O.K. – now can you – can you see that? Here's the one – it's only – its length's only a third as long as that one. Its width is only a third as long as that one so – if you fill in these . . . – Right, how many of the smaller ones can you fit in? – nine, right.

P: Is it that you square it – every time?

T: Yes – yes it is – why?

T: You think of it – cos you can . . .

T: It's difficult to – it's not easy to explain – here – this one fits in . . .

One striking feature is the ambiguity of the referents for the occurrences of the *it*s, particularly in the first pupil remark. A reasonable assumption is that the teacher was attempting to explain a general relationship, by means of a particular example, and that this general situation is the referent for the first pupil *it*. The second *it* presumably refers to the magnification factor (in

22

this case, three). The crucial expression in the above pupil formulation is *every time*, indicating that a generalization has been offered, even though the situation under discussion is quite specific.

In passing, note there is also an interesting ambiguity in the use of the term *length*, where it is used to refer both to general one-dimensional extent (contrasting with area or volume) but also to lengths in a particular orientation (length as contrasted with width). There is also the question of which is being reduced ('by three', lines being 'a third as long') and which is being fitted into the other (nine smaller ones fit into the larger). Later in the lesson, the expression 'all the lengths, length and width' is used, which highlights this difficulty.

Another aspect of pupil mathematical talk is the number of immediately-modified phrases. Many spoken formulations and revisions will often be required before an acceptable and stable expression can be agreed upon by all participants. Stability and confidence in a particular attempt or version is often indicated by an ability to repeat accurately what has just been said. Yet it is also the reported experience of many teachers that merely as a result of asking pupils to try to articulate what the difficulty is that they are experiencing, half-way through the resulting explanation pupils often say something like: 'Oh, I see now. Thank you very much for helping me.'

It seems that the act of attempting to express their thoughts aloud in words has helped pupils to clarify and organize the thoughts themselves. This can be true even of a pupil simply reading the question aloud, particularly if that pupil is young or a poor reader. Switching from the reading to the listening channel may activate different mental processes, or it may have more to do directly with the act of reading *aloud* rather than silently. I shall explore the importance of the role of an active listener later in this chapter.

Within the educational context of a mathematics classroom there are two main reasons for pupils talking, namely talking to communicate with others and talking for themselves. There is also a further justification, namely for the teacher to gain access to and insight into the ways of thinking of the pupils. *Talking for others*, in an attempt to make someone else understand something or to pass on some piece of information, is one of the many communicative functions which spoken language permits.

Talking for themselves involves situations where pupils may be talking aloud, but where the main effect is not so much to communicate with others as to help organize their own thoughts. (Barnes (1976) offers similar categories which he terms *explanatory* and *exploratory* talk respectively.)

Thus a second function which language permits besides direct communication with others is reflection on one's own thinking. Articulating aspects of a situation can help the speaker to clarify thoughts and meanings, and hence to achieve a greater understanding. By talking, thoughts are externalized to a considerable extent, which makes them more readily accessible to the speaker's own and other people's observations. The presence of another person may encourage reflection by the speaker on what has been said, or even provide an excuse for talking aloud, so there is considerable interaction between the two main reasons outlined above. No given excerpt is likely to be solely one or the other. In the next section, I look more closely at these two main reasons for pupil talk, namely talking for themselves and talking for others.

1 Talking for myself

As one extreme instance of talking for myself, repetitive talk – perhaps reciting the statement of a problem over and over – can help to clarify or fix a mental image, access to which is necessary for a solution. I personally find it difficult to 'hold still' certain geometric images, as they seem to suffer from a radar-screen wiping phenomenon, in which the image fades and then needs to be refreshed. Describing the image aloud, in words, assists me in holding it still so that I can then operate upon it. If I am engaged in performing an intricate computation, I also tend to 'talk myself through it'. While some of this self-talk is subvocal, I can find myself talking aloud irrespective of the presence of another person.

However, there are social and school conventions which militate against such vocalizations (e.g. working silently so as not to disturb a neighbour). Unfortunately, discouraging individual self-muttering can also serve to undervalue such an inner, exploratory or guiding monologue which may well be going on inside the pupil's head. Some pupils may not even be aware that they should be striving for such a monitoring of their mental activities, or that this has anything to do with mathematics. A useful technique for a teacher can be to deliberately 'turn up the volume' of their own self-talk (for instance, when explaining how to set about a particular problem) in order to allow pupils to eavesdrop on how a mathematically-sophisticated person operates.

There is a world of difference between tacit and externalized knowledge. One force of talking aloud is that it requires the use of words, whereas merely thinking to oneself allows words to be bypassed. It may be only when you discover a difficulty in expressing what you want to say, that you realize that things are not quite as you thought. Articulation can aid the process of reflection by affording better access to thought itself. In passing, note that this is also one of the central justifications offered by Papert (1980, p. 145) for pupils working with microcomputers.

> The computer allows or obliges the child to externalize expectations. When the intuition is translated into a program, it becomes more obtrusive and accessible to reflection.

What sort of talk occurs when pupils are working without a teacher present? This is one situation where talk to formulate one's ideas can be found. In the next extract (transcribed from the Open University videotape *Secondary Mathematics: Classroom Practice*), a pair of thirteen-year-old boys are attempting to work out a solution to the following problem.

Painted cube

Paint the outside of a cube red. Divide it into a number of smaller equal-sized cubes by means of various numbers of cuts. For each number of cuts, how many cubes will have 0, 1, 2, . . . red faces?

They have a physical cube in front of them which they are manipulating intently. The particular cube, part of standard multi-base equipment, is scored on the outside showing sixteen squares on each face. It provides a focus for attention and is supporting their speech in allowing a degree of vagueness which might otherwise require more refinement. The presence of the other provides a reason for talking aloud about the problem.

R: Right, look, look, there's eight two-sided on each cube isn't there. (Points to the two middle edge cubes on each edge of the top face.)

S: (Nods)

R: And then there's six sides. . . . So times six.

S: Except you've already counted that one on that side, haven't you. (Points to one edge that has been double counted.)

R: Oh yeah, yeah, I see what you mean. 2, 4, . . . 2, 4, 6, 8. (Counts the ones on the top face.) So there's only 16 of them. 2, 4, 6, 8, . . . (picks up cube and inverts it) 10, 12,

S: (Pointing to a vertical edge) There's some there.

R: How many? Eight? Which side? Which side?

S: (Takes cube) There's eight on there (pointing to the top face).

R: Yes.

S: (Turns it upside down) And eight on there.

R: Yes.

S: (Starts to turn to a third orientation) And there's . . .

R: (Grabs cube back) Keep it all on one side. There's eight on top there, aren't there. (Recapping his own original argument for himself?) . . . and eight there (cube upside down) and eight there (unsure)? . . . (raises hand) Sir?

While they are listening in part to each other, pupil R seems more intent on articulating *his* own perception of what is going on, rather than letting S explain his to him.

It is possible for a tape-recorder to listen less obtrusively than can any teacher. The presence of a teacher often distorts the communicative situation in a small-group setting, because pupils often feel that different rules apply; for instance, all of their comments are likely to be addressed to the teacher, and formerly active discussants may well wait passively to be asked questions. Also, various monitoring functions which the group may themselves have performed, such as deciding whether the speaker is correct or whether everyone has understood, tend to be *de facto* devolved onto the teacher.

If the talk is all focused on or channelled through the teacher, there can also be a difficulty with pupils vying for the teacher's attention, with everyone trying to talk at once rather than listening to each other. The result can be a number of individual conversations with the teacher. In order to diminish the teacher as the hub of the conversation, various techniques of deflection are possible. In the next chapter, I will give an example where a teacher steadfastly refuses to accept this role and regularly deflects requests for evaluation, for instance, over to someone else or back onto the group as a whole.

Another possibility for contending with this disruptive effect is for the teacher to alter the way in which a group is approached. Instead of arriving and requesting a summary of what has transpired in the group, an alternative approach can be to stand back on occasion, merely listening, attempting to be as unobtrusive as possible, before deciding that the time is right to initiate a dialogue. Pursuing individual pupil articulations and encouraging some pupils to help another to say something more clearly in a large teacher-led setting can also help pupils to express things for themselves.

2 Talking for others

The most familiar situation of pupils talking in a classroom is in response to a teacher-initiated question, and the response is then evaluated. This framework, namely the sequence *initiation* (I) – *response* (R) – *feedback* (F) (due to Sinclair and Coulthard, 1975), is widely applicable to much classroom talk. Below is a short illustrative example (to which I have applied their categories of interaction) from a secondary mathematics classroom. This transcript is taken from Yates (1978) and an extended extract (including this portion) is discussed from a different perspective in Chapter 3.

The teacher (T; P is any pupil response) has proposed to the class the problem of finding a means of communicating what is on the blackboard (a route map of major cities and motorway links in England) to someone in the next room. (I)

P: Morse Code. (R)
T: Morse Code, well that is not necessary. We can speak to him – he is only the other side of the door. If I was to put you on the other side of the door, you could hear what I was saying. (F)

P: Coordinates. (R)

T: Coordinates would be one way of doing it. That would be a very good way of doing it. What do you mean by coordinates? (F then I)

P: Say five across and down this way. (R)

T: Well that is a very good idea, it is one I had certainly not thought of. Any other bright ideas? (F then I)

P: Hold up a mirror. (R)

T: Hold up a mirror – it cannot go through a solid door. (F)

One aspect of this form of interaction is that the teacher retains control of the conversation, and another is that renewed initiation is not always required. In the extract provided, the pupils seem to be aware that a negative evaluation of a particular response requires further suggestions. Interestingly, the teacher's positive reaction to the suggestion of coordinates required him to request further offers, suggesting that although valid, coordinates did not form the solution he had in mind.

With such teacher-initiated pupil talk, the 'other' (for whom the pupil is talking) is the teacher, and the pupil may or may not have something he or she wishes to express. Focusing on the pupil responses in this extract reveals how brief and vague they were (often single words) when compared with the extent of the teacher's remarks. The motivation for talking is external, and speaking publicly can exert considerable pressures, particularly if pupils are called on individually by name. I have tried to show by means of the transcripts in the rest of this chapter that this teacher-controlled interactive style is not the only role for pupil talk when a teacher is present.

In the class excerpt given below, five thirteen-year-old pupils are working with their teacher on a peg-board game called *Leapfrog*. Two colours of pegs (denoted B and W, the same number of each) have been laid out as shown. In the diagram, there are three adjacent pegs of each sort and the asterisk indicates a single empty hole.

B B B * W W W

The aim is to exchange the positions of the B and W pegs where the moves follow certain rules. (These are that B pegs can move only to the right, by jumping a *single* peg of either colour, or by sliding into the empty space, and the w pegs can move only to the left, subject to equivalent rules.) The pupils were engaged in trying to predict the *minimum* number of moves required to interchange the positions of all the B and W pegs, for various numbers of pegs. Towards the end of their investigation, they tabulated the following results.

1 a-side	3
2 a-side	8
3 a-side	15
4 a-side	24
5 a-side	35
6 a-side	48
7 a-side	63

One of the pupils (Karen) noticed that:

$$1 \times 3 = 3,$$
$$2 \times 4 = 8,$$
$$3 \times 5 = 15,$$
$$\cdot$$
$$\cdot$$
$$\cdot$$
$$7 \times 9 = 63.$$

The teacher invited her to explain to the others what she had just noticed. She tried to *say* the pattern she had seen as 'the one on the right is always two bigger than the one on the left.' The teacher (T), trying for a broader generalization, posed a new problem involving a hypothetical man. (In the transcript below, K(aren), S(tephen) and C(aroline) refer to different pupils.)

T: He's thinking of a number of pegs a-side. He wants to work out how many moves it is going to take him.
K: What number is he going to choose first?
T: I'm not going to tell you. I've got a number in my mind and I want to know how many moves it's going to take me before I start off.

Notice the switch from *he* to *I* in 'I'm not going to tell you.' Assuming that the teacher is not being perverse, he can be presumed to be withholding the information for a purpose. Is his purpose clear to the pupils?

> K: Well, shall I just choose any number?
> T: No, . . . well, OK, to start off.

This last remark from the teacher signals that this is not really what he is after. The move from being able to operate successfully on any *particular* instance, to being able to encompass *all* such instances in a general statement, is a crucial one on the way to a successful control of algebra.

> K: Five a-side.
>   Five – add two on to five gives you seven.
>   Five times seven is thirty-five, which is how many moves.
> T: Now, I'm thinking of a number – I don't want to tell you what that number is. Will you try to explain to me how many moves it is going to take me?
> S: But I don't know what number you've got.
> T: I know. Tell me a sentence. I've got my eyes shut.

By closing his eyes, the teacher is trying to focus the pupils' attention solely on what is being *said*.

> C: Can I pick a number?
> T: No.
> S: Whatever number you've got in your head.
> T: Yes.
> S: You plus two . . .
> T: I plus 2.
> S: Add 2, and then you times it.
> T: Times what?
> S: The number what you started off with and the number that you've plus two'd.
> T: Do you think you could write that down for me?
> S: Yes sir . . . I can't write it.

Apart from the intriguing final remark, which will be explored further in Chapter 5 which is on aspects of pupil's written mathematics, there are many features of note in this exchange. A first general aspect is the marked preference of the entire group to work with particular examples, to *exemplify* the perceived pattern rather than to *express* it in its generality. The teacher's refusal to accept the complaint 'But I don't know what number you've got' was intended to suggest that the problem he had posed could be solved without knowing it. The phrase *whatever number you've got in your head* later becomes established within the group as a means for referring to the unknown. It is used not only to capture the uncertainty, but also to permit operations to be done upon the unknown number, treating it as an object with which it is possible to work.

Apart from an alteration from 'you plus two' to 'add two', a change possibly cued by the teacher's intonation when repeating what the pupil had said (such prompting by echoing being a common teacher strategy for encouraging further expression, and one that is discussed in the next chapter), Stephen's articulation of the pattern was initially:

you plus two and then you times it.

Once something is expressed, however haltingly and incompletely, then questions can be asked about the current formulation in order to encourage greater refinement, precision and clarity.

Spontaneous mathematical talk in response to a situation frequently involves indefinite words such as *thing*, as well as pronouns, particularly *it*. The definiteness of referents in mathematics is of considerable importance – in part, because their abstractness leaves greater scope for ambiguity and misunderstanding. Inquiring about the denotation of pronouns (usually *it*) is often the best way to start exploring the adequacy of some general expression; for instance, by requesting information about the various referents. The expression *you plus two and then you times it*, as it stands, contains neither what to start calculating on, nor how to interpret what is obtained at the end. For a spoken expression to stand alone, *disembodied* (that is, independent of its context for the necessary referents), all such components need to be included.

In terms of a developing dialogue, however, all the participants frequently know the subject of the conversation, and it would be artificial to pretend otherwise. This situation can place contradictory demands on a teacher. One general way in which a teacher's presence can interfere with developing pupil talk is by over-controlling it. The teacher may be too concerned wih the *form* of what is being said, at the expense of the *meaning* which the pupil is trying to convey. On the other hand, if pupils are to become aware of the characteristics of disembodied speech, then considerable work needs to be done to encourage them to modify and expand their initial attempts. One particular item in need of attention is the sense of self-sufficiency in the expression or explanation which allows it to stand independently of the foregoing discussion. How to contend with this tension may be one of the central dilemmas of communication facing teachers of mathematics.

The next transcript illustrates this difficulty. The teacher is attempting to work on the articulation of her pupils as well as undertaking some training in listening skills. The class is a group of low-attaining fourth-year secondary pupils and the transcript has been made from the Open University videotape *Working Mathematically with Low Attainers* (1985). The problem that the group is working on is to find the number of bricks necessary to build the fifteenth model in an ordered sequence, having been shown the first three in the sequence (see diagram). They are trying to justify why there will be twenty-nine bricks along the bottom of the fifteenth model by finding a general rule covering all the examples. (As usual, T is the teacher, P labels any pupil remark, J(enny) refers to a particular pupil.)

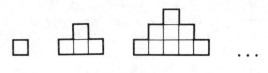

Diagram 1

T: They keep saying it and you are not listening – she keeps saying 'take away 1'.

J: I've got it – it is right. 15 add 15 is 30, right – take away 1 is 29 – that's the other answer we got.

T: Does it work with everything?

J: It will, it will.

T: How are you going to find out?

P: Do another one like it.

T: (To J.) What did you do just then?

J: 15, right.

T: Right.

J: Add 15.

T: Yes.

J: Equals 30 right – take away 1.

T: Concentrate, she's got something! Does it work on any more? Where does the 15 come from?

J: Cos that's the number you gave us last time, isn't it.

P: 15, that's what we had, 15.

P: You said the fifteenth . . .

T: Right, the fifteenth thingy.

J: The last number was 29.

T: And the last number was 29 what?

P: Take 1 from 30.

T: The fifteenth is the model number?

P: Yes.

T: The 29 is . . . what?

P: Is the bricks along the bottom.

T: The bricks along the bottom, which is what you are trying to work out.

P: Yes.

T: So now let Jenny explain to everyone what you do.

J: Right –

P: (Cuts in) Double it and take one.

P: (Cuts in) Add it and take one.

J: (Plaintively) They all know.

T: Double what and take 1?

P: The number that's there.

T: The number of what that's there?

P: (Inaudible)

T: I want you to write it down, now you've all had a go at saying it to me.

J: What, 'it works'? (Laughter)

. . .

T: (To Jenny) You're not telling everyone what to write down, are you? Let them write their own down. You write yours down, and let Marion write hers down.

This teacher thought it worthwhile trying to have Jenny attempt a clean, summary articulation of the general procedure. Yet it is hard to convey the purpose of being explicit about the referents in a situation where 'they all know', or of writing something individually about the group's discoveries.

Certain activities could be employed to encourage disembodied descriptions, that is, self-contained ones which can stand alone and be correctly interpreted by someone who was not present. One standard means for attempting to elicit such expanded accounts is to request an explanation as if for someone who had not been present, though it is seldom made clear to pupils why they should be interested in this activity. The above account raises questions about the perceived need among pupils for greater explicitness and precision with regard to spoken mathematical formulations. One general absence from many mathematics classrooms is much overt discussion of the *purposes* for the various activities and work that pupils are asked to carry out.

Constraints could be imposed on the context of the communication, either hypothetically or in reality, such as that the description has to be given over the telephone, or instructions provided for a blindfolded child trying to negotiate an obstacle course (Farnham, 1975), or even communication with someone in the next room. In an earlier example, I noted that the teacher said 'I've got my eyes shut' as a way of affecting the normal discourse situation, one in which pointing to written marks would remove the need to be sufficiently precise, *for the teacher's purposes*, using speech alone.

All these devices focus attention on the oral/aural channel and the self-sufficiency of the form of the speech itself in relation to the desired message to be conveyed. This is done by removing the possibility of visual support, and thereby altering those aspects of the environment which serve to make disembodied speech redundant. Below is an example of a situation which a teacher has set up in order to work on greater precision in speech about mathematical perceptions. The focus of attention for the whole class (twelve-year-old pupils in a middle school) was a poster of a great stellated dodecahedron, multi-coloured in a particular way (see cover), and they were invited to look at it in silence. A chair in front of the poster was designated the 'hot seat', where any pupil could opt to sit in order to describe something they had seen in this very complex picture.

In order to increase the verbal focus, the class rules for this

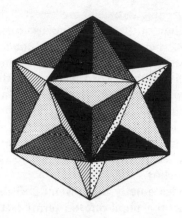

Diagram 2   Great Stellated Dodecahedron

activity included no pointing, touching or use of hands in any way, in order to indicate what was being described. Questions for clarification were to be directed to the incumbent in the 'hot seat'. This activity provides a particularly clear example of using words to point, that is to direct attention. The class have been operating in this way for about ten minutes when the excerpt begins. (T is the teacher, P labels any general pupil response and J(amie), those of a particular pupil.) This excerpt has also been transcribed from the videotape *Secondary Mathematics: Classroom Practice* (1986).

T: Does everybody see it as a three-dimensional object?

P: Yes (chorus).

T: (Invites Jamie to go to the front.)

J: (Taking the hot seat) Well, in the middle, right in the middle, there's kind of a triangle that kind of points out towards you, and all the fa[ces], the sides of it – there's other triangles that have been kind of broken up from it – say if they was joined to it, and when – by the dark green and the yellow and the light green, the kind of medium green kind of long triangles – and if the ones outside it was moved inwards, they would join the inside one.

Ps: Yeah, Yeah.

(Fellow pupil agreement, signalling they believe they are following his description.)

T: Oh, so it's going *in* . . . there?
    (The teacher is trying to focus on what the three-dimensional perception is due to, and also whether the picture goes 'in' or 'out' there. The definiteness with which the teacher used the word *there*, even though she was not pointing with her hand, gives a clear impression that she has a specific location in mind.)

J: Yeah, in like a proper triangle, cos you've kind of a distorted triangle in the middle, innit, well it's not a triangle, it's kind of . . .

P: Pyramid (echo).

J: (Counts sides quietly to himself) 2, 3, 4, 5, . . .

T: (Picking up the pupil-offered term) Is that *pyramid*, is it going in or coming out? I can see . . .

P: You're looking at it from the top?

T: Yes.

P: The one in the middle, if you look, erm, where the – you sort of seem to be looking at it from the top – on top of it.

T: So it's coming *towards* you.

J: It would make kind of – if these triangles at the edges of it was joined into it, they'd make kind of a flat shape with a pyramid sticking out of it . . . in the middle.

T: Where's the – can you describe where the three points of the triangle you are describing are?

(The teacher is having genuine difficulty in seeing the 'distorted' triangle that is the basis for Jamie's reference system.)

J: Er . . . On yer . . . In the . . . In the top left one, just inside, there's a kind of yellow and dark green triangle, and at the base of that, in the middle, there's one point. See that? (He turns to class, whereas before he has been mainly facing the poster, in a spontaneous gesture intent on monitoring his audience's comprehension before going on.)

T: (Shakes head and laughs.)

J: No, in the top right one.

P: (Interrupts) Top right?

J: Top left, over there.
    (At this point, a finger on his stationary hand is twitching frantically.)

T: What colour are we looking at?

J: It's yellow and dark green, it's kind of a little triangle . . . half-coloured.

P: *Light* green? (Correcting, alternative suggestion offered.)

P: It's *dark* green from here.

T: Yellow?

J: (Firmly) It's yellow and dark green, in the top left.

T: Yes, what you mean blue, I can see a blue one.

J: No, it's the point, the vertex of the top right, top *left* of it – d'you see that – the angle of it?
(*Possibly* trying to switch into more technical mathematical talk with the use of the terms *vertex* and *angle*, perhaps to see if that is the cause of the problem.)

T: Yes.

J: There's a point of a triangle, a yellow and dark green one there, the top point of it, if you was looking at it from here.

T: Oh, I see – you're seeing that as *one* triangle.
(Sudden revelation that what certainly the teacher – and perhaps many of the pupils – has been seeing as *two* adjacent, congruent triangles, one yellow and one dark green, Jamie has been seeing as *one* triangle, half-coloured yellow and dark-green. He needs to be able to talk about the middle of one of the sides of this triangle.)

J: Yeah, it's yellow and green, kind of, 'cept half-coloured.

T: Oh, I see.

J: And then at the bottom of that, in the middle . . .

T: Yes, I can see where you are now.

J: You can see one point – see that?

T: Point of what?
(In the complexity of the description and getting the group to see what he sees, the *purpose* of the description has been lost.)

J: The point of the triangle you told me to explain.

T: Oh yes (laughs).

The teacher commented afterwards that were she to do this activity again, she would ask fewer direct questions. She commented: 'Because *I* asked the questions, they tended to address me rather than the whole class as I had intended.' Part of the teacher's aim had been to use the poster as the focus of attention which would remove herself from centre-stage, and to encourage direct pupil–pupil communication without her acting as intermediary and, often, critical filter of what was said. The pupils were certainly not vying for the teacher's attention. They were listening and responding to each other, particularly to Jamie's attempts at communication. The structure of this activity, where the centre of attention is either on the poster itself or the incumbent in the hot seat, encourages such focused, active listening. (For a further account of this particular lesson, see Jaworski, 1985.)

In this activity, the teacher is working directly on communicative competence. The focus is on the meaning of what is being said and the successful communication of what has been seen. It is necessary to use language in a controlled way in order to be able to point, due to the constraints of the situation. One result may be to make pupils more aware that communication, particularly about mathematical ideas and perceptions, is difficult and should not be taken for granted. Particularly for pupils who are quick and fluent at computations, such a challenging task can serve as a timely reminder that not all mathematics is easy to routinize, and that struggle is likely, valued and supported.

This transcript also raises the question of what kind of talking is worthwhile in mathematics. Is more pupil talk, purely measured quantitatively, to be seen as a desirable end in its own right? Is pupil talk better seen as a vehicle to certain ends? And if so, is every type of pupil talk equally important, or should certain styles of talking be encouraged and others discouraged? It is to this very difficult topic that I turn in the next section.

3 What kind of talking is worthwhile in mathematics?

Brown (1982) has written an article exploring differences in style and purpose of classroom talk in an attempt to explain why, for her, merely increasing the amount that pupils talk in class is insufficient as an educational goal. She constrasts two different functions of spoken language, *message-oriented* and *listener-oriented* speech, and argues that the former needs to be explicitly taught in schools. In message-oriented speech, the speaker is goal-directed and wishes to express a particular message, to 'change the listener's state of knowledge' (p. 76) – it matters that the listener understands correctly. With listener-oriented speech, the primary aim is the 'establishment and maintenance of good social relations with the listener'.

While admitting that the distinction is simplistic and accepting that no conversation will use one or the other type of speech exclusively, she feels that both form and style of speech are clearly related to function. Mere exposure to others (i.e. teachers) who employ a different style is insufficient to enable pupils to acquire the more structured style of message-oriented speech. I feel that if pupils are to become proficient as *active* listeners, then they need to have access to the acceptable forms, as well as to know and value the purposes of using them. Brown provides a number of characteristics of listener-oriented speech, although she does not claim that her list is exhaustive. In the main, this form of speech is very general and non-specific. Among the particular syntactic attributes that she suggests are the following.

> Rate of delivery is slow, broken into chunks with a good deal of pausing.
> It is structured so that only one thing is said about a referent at a time. 'It's a biggish cat . . . tabby . . . with torn ears.' The relationship between the chunks is rarely syntactically marked. The speaker assumes the hearer will work out the conversational relevance.
> In talk about the immediate environment, the speaker may rely on physical context, e.g. gaze direction, to supply a referent: looking at the rain 'frightful . . . isn't *it*'.
> The speaker uses many non-specific demonstrative expressions e.g. 'I haven't done *this* before'.
> The speaker may fail to make it clear precisely which of several referents he is referring to – partly because he is using a lot of pronouns or very general lexical items. (p. 78)

The above features which I have selected characterize much in the pupil responses and attempts at explanation in the previous transcripts, in particular those of Jamie. He is using a predominantly listener-oriented style in a context which may have been helped by a message orientation. Therefore, one purpose of such classroom activities can be rephrased in terms of bringing about a greater awareness of the effectiveness of message-oriented speech in such circumstances, in particular an increased knowledge of the level of verbal specificity required for communication to stand on its own.

In mathematics, Brown's 'immediate environment' might mean written symbols, diagrams on paper, or even pointing at shapes sketched in the air. The contrasting form of message-oriented speech, Brown suggests, involves a more efficient packing of information, a more structured delivery with more syntactic marking, for example, involving words such as *because*, *therefore*, *consequently*, etc., together with more specific vocabulary. Jamie's use of the terms *vertex* and *angle* might have been examples of this last aspect in the dodecahedron extract.

Brown believes that all pupils are fluent in listener-oriented speech, but that message-oriented speech needs to be overtly practised. The characteristic linguistic forms by themselves are neither sufficient, nor a goal in themselves; they are a means to an end, namely clearer, more precise expression. I mentioned above some possible mathematical activities which could be employed to encourage such attention to precision in what is being said, as well as indicating potential situations to justify it.

It is not the case that only pupils use listener-oriented speech. Adults learning mathematics find exactly the same difficulties in being explicit. Brown also comments that 'shared knowledge allows inexplicit language' and I feel that mathematics suffers particularly from being seen as a subject where there is little point telling the teacher what she already knows. This merely restates a major difficulty facing teachers of mathematics, where the purposes behind a teacher asking for something to be done (e.g. try saying it out loud to your neighbour) are not clear to the pupils. If the activity is not seen as a reasonable requirement of the overt situation itself, many pupils will not see the point and so any potential benefits from the exercise will probably be lost.

Mental mathematics

This sub-section heading refers to a range of activities which involve working with mathematical ideas mentally, without recourse either to physical objects or to written representations. The most common mental mathematical activity in schools used to be mental arithmetic, an activity which has been much denigrated for the past twenty years, opinion changing only relatively recently. This denigration arose in part from a change in attitude towards the social atmosphere in which such sessions were commonly conducted, namely a public, competitive and time-pressured one, where all the emphasis was on rapid and accurate recall. It also arose from the fact that it was merely computational arithmetic which was the focus of the activity.

However, with a different emphasis and focus, working mentally can direct the pupil's attention to the inner realm of mental experience, where the root of mathematics itself lies. By encouraging pupils to exchange and explore their imagery, a teacher can find material to support a wide range of discussions on, for example, different representations, processes and methods. The emphasis on *personal* mental constructions means that the discussion will not appear redundant (i.e. the purpose *will* be clear), as the teacher *cannot* know what the pupil has in mind.

The context need not be arithmetic. Below is a brief account of a mental geometry activity and the discussion it provoked, as presented in the ATM text *Geometric Images* (Beeney *et al.*, 1982, pp. 26–27). (C is the teacher, P labels any pupil remark.) The class has been attempting to control mental images of circles and squares in response to descriptions by the teacher.

C: Close your eyes . . . think of a square . . . now let your circle sit on the square . . . now let it roll and focus your attention on the centre of the circle . . . what path does that centre take?

The corners caused disagreement. C asked for descriptions.

P: Well it goes parallel with the sides, then at the corners it sort of cuts a bit off.

C: Think of a bigger square round the first square, now cut a triangle off each corner.

P: No, it bends.

C: What do you mean?

P: It turns, like a bend.

C: Which way?

P: Depends on the corner.

C: How much does it turn?

P: Er . . . a quarter at each corner . . . like quarter circles.

The emphasis in this extract is properly on communication about a personal experience to others who are attempting a

similar activity. The central purpose of speaking in such a context
is to convey what has been *seen*, or perhaps only glimpsed, in a
mental realm 'two inches behind the eyes'. Access for others is in
the main through words, the control of which can conjure images
for others. The teacher is asking questions because he actually
wants to know what they are seeing and cannot otherwise find
out. One of the continually surprising aspects of mathematics is
that such negotiated agreements about individuals' mental images
are not only possible but are regularly achieved.

The extracts in this chapter have raised important issues about
the functions of classroom talk. In particular, they perhaps
suggest that merely increasing the amount of pupil discourse in
mathematics classrooms may not, by itself, prove beneficial. This
is a very important point; I believe pupil talk *per se* is not
necessarily a good thing. Some of the desired attributes for the
teacher to aim for are that pupil talk should be focused, explicit,
disembodied and message-oriented. Jamie's discussion, although
much of it was highly listener-oriented in style, was focused on
the task and displayed a considerable sense of audience.

If a teacher is working towards these goals, both the goals and
the reasons and justifications for them should be shared with the
pupils as much as is possible, so that the pupils too may realize
why they are being invited to do things which the situation itself
may not seem to require. To reiterate my earlier concern, this
working towards such goals should be tempered by an awareness
of two interfering factors. Firstly, many communicative situations
do not require, *in themselves*, such attention to precision and
explicitness in order to be reasonably successful. Secondly, over-
concern with the *form* of the talk, at the expense of the content
(the meaning that they are trying to convey), can well result in
pupils not bothering. It is finding a way to blend these conflicting
elements together, in conjunction with the teacher's own
purposes and intentions, which provides a central challenge for
mathematics teachers.

4 Oral and aural proficiency

Pupils need to be trained to be active listeners. The teacher
cannot be everywhere at once, and so pupils need to be
encouraged to take on some of her 'critical' functions within
small groups. Brown (1982) made this comment on the role of

the listener in listener-oriented speech: 'the listener rarely interrupts to identify who or what is having what said about them – one's general impression is that the listener is often only doing a fairly sketchy interpretation' (p. 79).

For many pupils, mathematics is predominantly passive and aural, sitting listening to a teacher going on about triangles, decimal multiplication or algebraic manipulation. Although the teacher is operating with a message orientation, the pupils may be receiving what is being said with a listener orientation (maintenance of good social relations with the teacher may be of considerable importance to the pupils, so they feign interest in the proceedings). As suggested above, in a listener-oriented mode, it is uncommon to interrupt when what is being said is unclear. The social pressure conventions of a listener compound those of the accepted view of the pupil role to produce this reticence.

It is equally important for a teacher to attempt to increase the active listening skills of pupils, so that pupil–pupil discussions may be more than casual conversations. Pupils need to become aware that there are different ways of seeing situations, which may all have merit, and also that it is important to monitor the rest of the group to see whether or not what has been said is clear. Their role is not a passive one: rather, it involves an acceptance that being precise about the situation is important. In general, the skills of deciding what is relevant and how to refer to things are hard-won. However, mathematics teachers need to encourage both critical listening and verbal fluency in explanation.

Even now, it is uncommon for pupils to be asked *how* they attempted to solve a problem, rather than for the answer they obtained. They are provided with few opportunities to practise giving explanations and justifications, or even to express something which they themselves have perceived. All the arguments they see, whether written in a text or presented orally by a teacher, tend to be fairly polished and fluent, rather than full of hesitancy, stumblings, backtracking and revisions, as theirs would be. (Recall the Holt quotation given at the beginning of the chapter.) Teachers are very good at giving explanations – they also have a great deal of practice. In general, pupil explanations are rarely elicited, nor are members of the class encouraged to listen attentively to or try to amend or improve each other's contributions. It is more common to push on in search of an acceptable formulation, as was seen in the Yates extract in the first section.

It is also seen as part of the teacher's function to comment, judge and amend. So if a teacher desires to give up some of this critical role, perhaps for reasons of desiring to improve the oral fluency of her pupils, then this needs to be clearly and regularly signalled to the pupils themselves. The teacher can model for the pupils what being an active listener means by engaging with and encouraging their hesitant attempts at expressing what they see in various mathematical settings and situations. The whole of this chapter suggests that the oral proficiency of their pupils be taken as a serious goal by all mathematics teachers.

The role of examinations

There is very little emphasis on oral proficiency in mathematics in British schools. In certain European countries, such as West Germany and Denmark, there is a general tradition of oral examining (and hence appropriate instruction toward that end) in all subjects, including mathematics. The position in this country is almost entirely the reverse: the vast majority of mathematics assessment (and that of most other subjects), whether at the individual school or at the national level, is by means of individual, pencil-and-paper, timed examinations. Post-Cockcroft, with a renewed interest in alternative models for and modes of assessment, the possibilities for oral examining in mathematics are being reluctantly explored by some examination boards.

There are obvious drawbacks to oral examining (even in a non-individual context), including the facts that it is extremely time- and labour-intensive, and is considered less reliable than written examinations. It is my contention, however, that the reliability of mathematics marking is artificially high due to the narrow range of questions which are asked. Questions not meeting the existing high levels of mark/re-mark reliability are thrown out on the grounds of their being unfair. Mathematics examinations are heralded as masterpieces of certainty and exactitude of marking. In fact, this reliability (as so often is the case in educational research as well) has been acquired only by a radical truncation of the sorts of questions 'allowed'. Only very recently have examination boards been more willing to accept considerable responsibility for the influence they bring to bear on school syllabi, as well as what actually goes on in secondary classrooms. One external means of encouraging direct attention to oral fluency is by valuing it in terms of examination representation. A contrary view, at once idealistic and cynical, would be that given the destructive effect of content-based written examinations on the learning and teaching of mathematics in secondary schools, *nothing* which teachers value as being educational should be permitted to be included in public examinations, for very real fear of its destruction.

5 The role of discussion in mathematics

I mentioned in the Preface that the topic of *discussion* in mathematics is currently receiving a good deal of attention. As with investigation and practical work (two other 'missing' attributes from mathematics teaching), there have been considerable demarcation disputes about what is or is not to count as discussion. In addition, there seems to have been an assumption that discussion is (always and automatically) a '*Good Thing*' (Sellar and Yeatman, 1960).

Discussion has quickly become a new panacea, part of the new orthodoxy. The appeal for more discussion is part of an attempt to redress the balance of mathematical activities towards the oral, as well as signalling a desired move away from teacher exposition. In addition, although as a category to be made into a topic it is even more anomalous than an investigation, the expression *the discussion lesson* has been used (as one might talk of the fractions lesson or the algebra lesson). No matter what sort of entity it is, in teaching everything seems to end up as a topic for a lesson.

It is partly for these reasons, particularly the propensity for discussion to be mistaken for a goal in itself rather than a means to an end, that I have opted in this chapter to use the more neutral term *talk*. The presumption of the value of discussion becomes translated into questions about different forms and purposes of talk (as seen from both the pupils' and the teacher's perspective) in mathematics classes.

Calls for increased discussion are not new. They have been a regular feature of many reports on the teaching of mathematics over at least the last hundred years. It is of considerable interest to me that despite the apparently innocuous request to increase the amount of discussion in mathematics teaching, this has not been achieved. Such a situation seems worthy of closer consideration.

The Cockcroft report, as part of its recommendations for broadening the range of mathematical experiences which pupils encounter in class, called for discussion to play a wider role in the teaching and learning of mathematics: 'Mathematics teaching at all levels should include opportunities for . . . discussion between teacher and pupils and between pupils themselves' (para. 243). In a Schools Council document. *Mixed-Ability Teaching in Mathematics* (1977, p. 58), the authors noted: 'Many classrooms were characterized by their lack of discussion.'

Following these recommendations, there has been a resultant confusion bordering on incomprehension concerning *how* to talk mathematics. It is very difficult to glean from the Cockcroft report itself, or even the follow-up document *Mathematics 5–16* (HMSO, 1985), what is meant there by the term *discussion*. On the one hand, I feel it is unhelpful to elevate discussion to mean a pseudo-mediaeval disputation as some seem to envisage. On the other, the term is clearly intended to refer to a more specific category of verbal interchange (whether pupil to pupil or the teacher with an individual, a small group or a whole class) than does the all-encompassing term *talk*. The Jamie dodecahedron extract indicates pupil–class discussion as a further possibility. That situation certainly entails negotiation of meanings and sharing of points of view about mathematics which seem to me to be important component aspects of discussion. This difficulty in specifying the limits to discussion is one further reason why I have chosen to focus this chapter on talk rather than on discussion *per se*.

How, then, are situations to be engendered in mathematics classes where genuine discussion about mathematical topics may be achieved? Under the heading of mental mathematics, I earlier suggested one source of activities which, if appropriately handled, can result in worthwhile discussion and negotiation of meaning to a definite end. But this requires the teacher to be willing to let the class determine to some extent the content of the lesson, together with the direction in which it will move. This is one reason why the calls for increased discussion in mathematics are not necessarily as bland as they may seem at first. Interpreted in this way, such recommendations contain the seeds of a radically different way for a teacher to operate in a mathematics classroom.

Even if the value of genuine discussion were accepted by all teachers, perhaps on the grounds that it enabled them to discover a considerable amount about what their pupils thought, as well as encouraging more talk of particular kinds (and therefore benefits), there still remains the question of how to accomplish it. It is clear that trying to 'hold discussions' creates a number of difficulties. The *status quo* can be hard to alter, even by a teacher who has decided to try to introduce a more discursive atmosphere. Pupils' views and expectations of what *should* go on in mathematics lessons are often quite rigid. One pupil complained after working for a period, attempting to express the patterns inherent in a sequence of drawings: 'We all spoke too much – we haven't done any maths'. For him, it transpired, mathematics consisted of written 'sums' done in exercise books. Pupil conceptions of mathematics do influence what can happen in the mathematics classroom.

There is one further general difficulty with the notion of discussion in mathematics. Mathematics is not commonly viewed as a discursive subject. Surely, the argument runs, there are right and wrong answers to everything, together with clear-cut methods to be taught and learnt for finding them. So how can mathematics be discussed when there is no place for opinion, informed or otherwise? While there might be open problems at the frontiers of mathematics, it is all sorted out and written down at the school level. Although I have somewhat encapsulated this argument, I have tried not to caricature it.

This conception of mathematics is quite widespread, and whether or not it reflects an accurate picture of mathematics and mathematical activity, such teacher and pupil perceptions play a central role in determining how a subject is taught. Thom, a world-renowned mathematician, has claimed that, 'all mathematical pedagogy, even if scarcely coherent, rests on a philosophy of mathematics' (Thom, 1973, p. 204). The perceived problem with discussion provides a clear instance of Thom's point. Many teachers do not see the value or even the possibility of discussion in mathematics as a consequence of the view of mathematics which they hold.

Summary

The opening remarks of this chapter outlined two separate but linked justifications for encouraging verbal formulations of mathematical perceptions and ideas. There is a power in saying things aloud and frequently ideas can only be looked at properly when they are externalized to some extent. In addition, verbalizing externalizes pupils' thinking, rendering it public, thereby allowing teachers opportunities to discover their pupils' ideas and beliefs. It also allows pupils to check that they have understood what has been said. More importantly, eliciting mathematical talk focuses attention on argument and conviction by means of explanation, as well as on the task of finding more precise and succinct expressions which may therefore be more readily worked with and verified.

I do not think that merely increasing the amount of pupil talk in mathematics classes should be seen as an end in itself. It is essential that the conversations be task-focused and the style and level of explicitness of the talk are both important. The pupils need to have some idea of *why* they are being encouraged to talk, for among other things, without knowledge of the purpose, criteria cannot be brought into play about adequacy and explicitness of the account.

The question of for whom the talking is being done is of central importance. The pupils are guided by normal conversational expectations about relevance, explicitness, etc. The teacher, whose covert aims may include a desire for a greater oral fluency in her pupils, often requests things which the situation does not apparently require. Because teachers are often inexplicit about their purposes, they can end up at odds with their pupils. There is a considerable need for oral fluency in mathematics to be recognized overtly by pupils as an appropriate aim and for teachers to find ways to acknowledge and discuss their purposes with their classes.

All the above activities depend on pupils' verbal expression being seen as an important part of teaching and learning mathematics, yet little time in class is customarily devoted to such an end. Many subsequent surveys of 'teacher talking time' have replicated the results of Flanders (1970) who found that, on average, for two thirds of the time someone was talking in a classroom, and for two thirds of that time, it was the teacher who was talking. There is an obvious tendency for teachers to assume responsibility for and hence control over the verbal exchanges in the classroom, with a consequent over-eagerness to fill silences. This can result in domination and direction of the verbal interchanges. In the next chapter, I look at certain teacher strategies (linguistic and otherwise) which I term *gambits*, which determine to a large extent the pattern of communicative interaction in mathematics classrooms.

3

Overt and covert classroom communication

School was all talk, of course, but in a different way. Being told, not telling.

Penelope Lively, *The House in Norham Gardens*

There is a sense in which, in our culture, teaching is talking.

Michael Stubbs, *Language, Schools and Classrooms*

The aim of this chapter is to explore some of the types of interaction commonly found in mathematics classrooms and the sorts of communication they permit. In the context of chess, the term *gambit* refers to a move or a series of moves which involves a possible sacrifice on the part of the instigator, but which is intended to produce an overall advantage. This idea can be fruitfully applied to the teaching situation to describe, for example, the strategy of teachers inviting their pupils to converse in pairs (for varying lengths of time). One sacrifice involved is that, by encouraging talking in pairs, even if the teacher circulates to monitor and participate in some of the conversations, she rescinds control or even an awareness of many of the verbal interchanges going on in the class.

One potential gain as a result of this sacrifice is that subsequent whole-class discussions may well fare much better, as many more pupils may have something to contribute, having rehearsed its expression in the less threatening context of conversation with a neighbour. (This is just one of a number of discussion strategies which are explored in Mason and Pimm (1987) which contains a detailed examination of the nature and place of discussion in mathematics, together with an account of various classroom techniques which may help to bring it about.)

There is a possible misconception implicit in the use of the word *gambit*, concerned with consciousness. In talking of strategies or gambits, I do not intend to give the impression that I view teaching as a manipulative or completely conscious activity. I am well aware that many classroom decisions are made at an instinctive level, on the spot, in the heat of the immediate situation. Nonetheless, separating out certain gambits makes it possible to analyse teaching at a certain level, and to be more precise about a range of teaching options which are available in response to a particular situation. This may allow a teacher greater flexibility in the future. It can also help to single out and focus on particular aspects of classroom practice and bring them to conscious attention.

The notion of *sacrifice*, while it may not be appropriate in every case, serves as a reminder that many current practices may have unintended outcomes that ought to be thought about. For example, by not answering a direct question, perhaps deflecting it by asking other pupils what they think, or by inviting the questioner to explain what he has found out so far, a teacher relinquishes the opportunity to give a clear explanation of what she thinks is going on. Depending on the values and beliefs of the teacher, this may be seen as a greater or lesser sacrifice. A long-term advantage connected with this gambit may be that the class becomes less reliant on the teacher as a source of authority, and can move towards a greater personal mathematical judgment and autonomy.

A third example of a teaching gambit involves the use of silence. A silence, following the asking of a question for example, allows pupils thinking time, but means that the teacher will have to contend with any possible embarrassment she may feel. This may be strong if a teacher feels responsible for controlling communication, that if no one is speaking, it is up to her to initiate dialogue. A teacher who believes the gain is important will have to rescind some control over the spoken communication channel.

These particular gambits are, in a sense, non-verbal, though they have major implications for the level and type of pupil talk in the classroom. In the next two sections I concentrate mainly on spoken gambits. (One instance of this is *echoing*, which I mentioned briefly in the last chapter.) Underlying this general aim is an attempt to examine whether some of the strategies of mathematics teachers are acting as inhibitors of, or a deterrent to, pupil talk about mathematics. Certain styles of communication are very common, and I shall attempt to illustrate some of them by means of an analysis of transcript excerpts.

Reading classroom transcripts can have the feel of reading a playscript (perhaps written by Harold Pinter). Transcripts of actual classes regularly indicate little verbal interaction between pupils themselves (particularly about mathematics), and much of the class interaction is led by teacher questioning. I begin by examining some of the verbal strategies which teachers employ, whether consciously or unconsciously: in particular, I am concerned with how questions are asked, and to what end.

1 Ways of asking and answering questions

Rather than start with a genuine transcript, I have chosen a very familiar model for teaching style in mathematics, namely Plato's dialogue *Meno*. Much approval has been poured on the eponymous Socratic method of instruction, yet viewing the *Meno* dialogue as a classroom transcript provides evidence for disquiet. Below is a short extract from the dialogue. As you read it, focus on the type of response that the style of questioning allows.

Socrates: Tell me boy, do you know that a figure like this is a square?

Boy: I do.

Socrates: And you know that a square figure has these four lines equal?

Boy: Certainly.

Socrates: And these lines which I have drawn through the middle of the square are also equal?

Boy: Yes.

It continues in this manner for a number of pages.

The teacher domination and control of language is evident. The pupil is relegated to the position of, at best, acquiescer in what is happening. Socrates proclaimed 'Agree with me, if I seem to speak the truth.' That is precisely all the boy is allowed to do. (The social pressure operating in such a situation has led to this technique being dubbed 'proof by intimidation'.)

In reading, there is a widely-used (but appallingly-named) technique called the *cloze* procedure which is used, among other things, to assess text reading difficulty. Single words are deleted according to some general scheme (e.g. every fifth or seventh one) and the task involves pupils trying to fill in the missing words. Teacher questions in mathematics are frequently what might be called 'clozed', that is, where a one-word answer will suffice, and this is often all that is, in fact, desired. The response is almost always given without repetition of the teacher's sentence and seldom is encouragement given for a full sentence reply. Confirmation, on the teacher's part, is often signalled by repeating the offered word (a variation on a verbal gambit of echoing), as if the pupil had offered a prompt in a play, and then the monologue is continued as if the exchange had not occurred. For example, this short extract comes from the same secondary class of fourth-year pupils (working on the topic of the area of similar figures) as the first extract in the last chapter. (T is the teacher, P a pupil.)

T: Supposing you started with that shape and you increased its length by two, what would have happened to the area? It would simply have . . .
P: Doubled.
T: Doubled – but that's only moving in, increasing in one dimension – if you've got to increase the width by two as well – then you've got to double it again.
So – you would be doing – first of all times two to double the length – and then times two again – to double the width – so altogether you must have multiplied . . .
P: By four.
T: By four, by two squared . . .

A slight rise in intonation, coupled with a pause, cues the pupil that some response is called for. In the second exchange, attempting to bring out the structure of the general situation, the teacher accepted, signalled by repetition, then modified the pupil's answer. I use the word *answer*, although there is no syntactic marking of a question in anything the teacher has said. At what stage do pupils learn that this is a strange way teachers seem to have of asking questions? Strange, that is, if the aim is to get pupils talking about mathematics.

The 'clozed' style of questioning provides a good example of a spoken teacher gambit. One of the advantages is that it allows the teacher to maintain control of the discourse, while focusing attention on particular items along the way. It is also a device to break up a teacher monologue and acts as a check for the teacher that the particular pupil questioned has grasped what is being explained. This latter reason emerges from a deep-rooted belief on the part of many teachers that there is a power in someone saying things aloud, and therefore it is better for the pupils to say the central part for themelves, rather than merely hear it expressed by the teacher.

The sacrifices involved with this particular gambit, however, are at least two-fold. Firstly, the scope for possible topics for such questions is very narrow, as the answer can usually only be a single word which fits in with the grammatical structure already specified by the 'floating' part of the teacher's utterance. Secondly, the constraints of the form deny pupils practice in formulating whole sentences or longer explanations, particularly of *their* perception of the situation. There is the danger that the questioning may degenerate into guess-what-is-the-particular-word-in-my-mind. It is also possible that the teacher is so hooked on a particular word that he may overlook or even say 'no' to alternative, equally valid possibilities.

Below is a second extract (transcribed from an Open University EM 235 videotape) which illustrates a number of these features. The situation is an adult working with a group of nine- and ten-year-olds on the topic of area of rectangles and parallelograms. The question, 'What has changed, and what has stayed the same?' has been used with a previous pupil (Robert). As we join the conversation, another pupil (P) has been invited to explain what she has done and found out.

T: Now, you tell me about yours. You've got . . . you've got a parallelogram.
P: Yes.
T: Change it into another one.
P: (Moves one triangle to make a rectangle.)
T: You're going to change it into a rectangle. Now, you tell me what has changed and what has stayed the same.
P: This little piece just . . . changed.
T: Right, the overall *what* has changed?
P: The shape.
T: The shape has changed, and what has stayed the same?
. . . (slight pause)
Did you hear what Robert just said about his? Robert,

what did you just say about yours?
R: The er parallelogram stayed the same and so did the area –
    so it is just the shape has been moved.
T: Right, the shape has changed and the area stayed the
    same. Is that true of yours?
P: Yes.
T: You tell me then. You tell me what stayed the same and
    what changed.
P: The sides, the sides . . .
T: The *shape* changed.
P: Yeah.
T: The *shape* . . .
P: Yeah.
T: You tell it to me.
P: The shape changed.
T: And what stayed the same?
P: The area – in the middle.
T: The area, that's right.

One difficulty peculiar to the teaching situation occurs when
teachers ask questions: not to discover something they did not
know, but in an attempt to ascertain whether or not the pupil
asked knows it. The existence and predominance of this second
purpose for questioning in school can result in some peculiar
exchanges. A history teacher, who had forgotten, asked, 'How
many wives did Henry IV have?'. A diffident pupil hesitantly
replied, 'Er, was it three, Miss?'. The teacher replied in an
annoyed tone, 'I don't know. Why do you think I'm asking?'.
Possibly as a result of years of the examining sort of questioning,
pupils can often attribute a form of omniscience to their teachers.
Hence a genuine request for information, possibly about how
they did something, is rendered either redundant or threatening,
thereby distorting the normal communicative context.

Answering questions from pupils

There are numerous ways of reacting to a question asked by a
pupil in class. Obviously, the response will depend in part on the
content of the question and why the teacher believes it has been
asked (e.g. to deflect the teacher from the current task or topic).
Answering the question directly is one option, but by no means

the only one. I mentioned earlier the possibility of deflecting questions. This is one gambit which can allow a teacher to escape from the tyranny of the I(nitiation)–R(esponse)–F(eedback) framework which I outlined in the last chapter. Tyranny, because it locks the teacher into 'centre stage', acting as controller of the communication, as well as heavily influencing the *types* and range of spoken pupil contributions in class. It is likely to have been the format of so much of pupil experience in school, that to attempt to operate in a substantially different mode can involve considerable time and effort.

The gambit of feigning ignorance is often used by mathematics teachers working in an investigative context, as a device to deflect a direct request for information, guidance or evaluation. It often happens, however, that the teacher is rightly disbelieved by the pupils, which can then cast doubt on whether the teacher is being honest with them in other respects. Because of difficulties of this sort, some teachers prefer to seek out different means to achieve the same ends.

Another way to diminish direct questioning is for the teacher to attempt to remove herself from centre stage (e.g. by focusing class attention on a poster or a microcomputer screen). A further way is to refuse (overtly or covertly) to respond. If it is overt, e.g. 'I'm not going to tell you', then some reason should be provided – otherwise the teacher runs the risk of being thought awkward or perverse. This provides another instance where the purposes for doing something apparently unusual needs to be made clear.

The next transcript is taken from a lesson where the entire group (a first-year mixed-ability secondary class containing thirty-two pupils) together with their teacher were seated on chairs in a circle. The topic of the lesson was modular arithmetic (base five) and the opening activity was firstly for the pupils to number themselves in turn round the circle and then for each to take a Cuisenaire rod (the first five rods were available in large quantity), one by one, in the same order. The rod to be taken was determined by the pupil's number in the circle. The first person took the shortest rod and the second the next shortest and so on. The sixth person in the circle took the same rod as the first person did, namely the shortest (white) rod. This same cycling back occurred every five. The two main challenges offered by the teacher were firstly to ascertain the matches of colour to number (e.g. what colour rod will number thirty-seven in the circle be, or can we work out who has a red rod?) and then to do arithmetic on the colours (e.g. to find a way to add a red and a yellow).

My particular interest here is in the teacher's technique of dealing with pupil responses to her questions. When the extract begins, the pupils are coming out to the middle, taking their rods one by one. There are a number of pupils who have yet to take their rods before it will be Debbie's turn. (T is the teacher and D(avid) a particular pupil.)

T: Stop a minute. What colour is Debbie going to take? . . . (pause) David, did you want to say something?

D: Yes, cos every fifth one from William is going to be a white, and every fifth one from the next person on is going to be a red, and every fifth one from the next person is going to be a green, eh? (The last noise being a questioning intonation possibly asking for confirmation.) (General confusion and discussion of this announcement.)

T: Is that right? Is what David's saying right?

P: (Partial chorus of yesses.)

T: Right, let's take the rods then quickly, so we can find out. Can you think what colour you are going to be and then come and get them. . . . Right, who's like me, who's got a red one? Put up your hands if you've got a red one. . . . Julie?

J: Seventeen.

T: Seventeen's going to be red. David?

D: Twenty-two and twenty-seven and thirty er two.

T: (To the class) Is that right? . . . (To a particular pupil) Are you thirty-two?

P: Yes.

T: What colour are you?

P: Red.

T: Red.

P: Miss, it's all the numbers that end with two or seven, that have the end in two or seven.

T: It's all the numbers that end in two or seven.

Same P: Yes.

T: Will be red.

P: Are going to be red. Green would be three and eight.

T: (To the class) Can we work out any numbers?

(If you have access to the videotape *Secondary Mathematics: Classroom Practice* (1986) from which I have transcribed this excerpt, I strongly recommend that you watch it, as so much of the quality of this lesson is untranscribable.) One aspect which cannot be adequately conveyed by a transcript alone is the friendly but neutral voice tone in which many of the teacher's remarks were made, reflecting her refusal to evaluate the responses that were offered to her questions. The activity was carefully designed so that many of the pupil conjectures were able to be checked by examination or experimentation.

Echoing, in the sense of repeating a version of a pupil comment, is a device which allows this teacher a turn in the conversation that returns the conversational impetus back to the group if someone has anything further or new to add. She deflects onto the group as a whole the responsibility for evaluation of a contribution arising from one of her questions if echoing does not successfully contend with this or if she wants the group to think particularly about a comment. The teacher being *in* the circle helped somewhat to defocus attention from her and onto the activity. The circle also meant that everyone could see and talk to everyone else, thereby encouraging pupils to address comments to each other and allowed small pockets of local discussion to develop on occasion. It was also an essentially large-scale activity which required the participation and involvement of a big group.

I have only scratched the surface of the general and important topic of questions in mathematics teaching and exemplified a couple of gambits observable in the asking and answering of questions. As an activity, I invited a group of teachers to make tape-recordings of themselves either with their whole class or with an individual pupil, and then to transcribe them. Some of the general things they noticed included how much, both relatively and absolutely, they talked: seeing the relative contributions mount up on a page made quite a considerable impact. In addition, they remarked on the paucity of pupil responses and how short they were, often comprising only one or two words. One teacher felt she was not providing the space or opportunity for longer answers by the way she actually formulated her questions.

If teachers become more aware of what they do and some of the consequences (intended and unintended) these various gambits have, they are in a better position to work on their own teaching and undertake explorations of what can be achieved. In the next section, I move away from questioning in particular, to look at different aspects of classroom interaction.

2 Further aspects of mathematical classroom discourse

One of the distinctive features of discourse about mathematics is the widespread use of a technical vocabulary. Mathematicians have developed an accepted public language with which to communicate to others, which has evolved primarily to meet the needs of expert users. The existence of this approved manner of communicating affects the spoken environment of mathematics classrooms by influencing the language in which teachers feel they *should* communicate. In particular, this can involve teachers in the (surreptitious) correction of (or failure to build on) the unconventional but sometimes strikingly evocative language of their pupils. Certainly, one of the widely perceived tasks of teachers is to shape their pupils' mathematical usage towards the approved 'dialect'.

The next extract involves a teacher with a second year secondary class and the topic of pie charts. (T, the teacher and P any pupil.)

T: What do you think a pie chart is?
P: It's a circle.
P: It's 360° at the middle . . . and you measure the degrees.
P: It's divided into sections, and it's the size to indicate a piece of information.
T: The size depends on the information we are given. Now can anyone think very carefully why we might want to use a pie chart rather than a bar chart or a pictogram?
Class: No.
P: To show how many percent there are.
T: Well, sort of percentage. I don't want you to go into details.
Same pupil: So you see how big it is . . . takes up the space.
T: You're on the right lines . . . what I'm looking for. Pie charts are used mainly for comparison. What I want you to do is put the heading 'Pie Charts' in your books.
P: Which book, Miss?

T: Put the date – it's the 23rd today, isn't it?
P: 22nd.
T: So you can write 'A pie chart is a circle divided into sections' since that's your word, we'll write down what you gave me. From now on, we'll talk about sectors, not sections. Perhaps you could write that another word for a section is a sector.
P: What, Miss?
T: It's divided into sections – you see that? Another word for a section is a sector.
P: How do you spell *sector*?
T: Right, so another word for a section is a sector. You can write 'and the size of the sector depends on the information we are given'. Do you all understand so far? Put your hand up if you don't. 'Pie charts are used to compare the relationship between the different items' (Writes on board). Do you understand if I say 'between the different items'? Normally we'd put a label in here. We could say 'different things'. That doesn't sound very mathematical. It's what you usually say. I'll call them items.

There are a number of interesting exchanges in this transcript. The desire on the teacher's part to get the official term *sector* into play seems to negate the earlier claim about using the terminology of the pupils. 'That doesn't sound very mathematical' is another indication of this teacher's concern with the 'official' language. The conventional nature of this terminology is suggested by the sentence 'It's what you usually say', and is followed by an acceptance of this convention. No indication was given to the pupils of why *sector* might be more acceptable or 'mathematical' than *section*. This extract quietly attests to the unequal distribution of authority and power over even the language to be used in a class setting, with the teacher acting as arbiter of acceptability, reformulating and correcting without comment in certain instances.

Stubbs (1983b, p. 102), summarizing the work of Barnes (1969) on classroom language, writes:

A teacher may see the language of his subject as having an intellectual function of allowing concepts to be precisely expressed. But the teacher's language will also have a *sociocultural function* of supporting his or her role as teacher. And, from the pupils' point of view, each new term may have a predominantly sociocultural function: it is 'the sort of thing my teacher says'. Barnes is here pointing to a source of *sociolinguistic interference* between pupils and teachers who have different notions of stylistic conventions. That is, a teacher may use a certain style of language, not because it is necessary for expressing certain ideas, but because it is conventional to use it. But the pupils are unlikely to share these conventions.

The above instance of *sector* versus *section* seems to be a clear example of this phenomenon. However, the existence of this official language in mathematics and the concern of many teachers to require its use by their pupils, form some of the many constraints on genuine discussion, which I pointed to in the previous chapter. In a geometric context, use of the technical term *sector* might serve to mark a necessary distinction, for example, in needing to distinguish a *sector* from a *segment* of a circle. In such a circumstance, the word *section* would not be sufficiently precise for the teacher's purposes. But as in so many cases, this presents another dilemma for the teacher, in that the pupils are likely to be immersed in the demands of the here-and-now, while the teacher also has to be thinking to the future and the demands that upcoming work is likely to impose.

One extreme alternative to the above overruling of the pupil-generated terminology is to permit or actively encourage what might be called an *esoteric* classroom. In other words, a teacher could set out to create with her class a shared language and set of meanings which operate within that classroom, but which may not have a wider currency. Handled with care, as well as overt discussion of the role of language and written notations, a completely different perception of mathematics and the way it proceeds might ensue.

Modification of the outcome of a lesson by what has been said

I suggested briefly at the end of the last chapter that such a modification provides one signal that genuine discussion is taking place. Another is that modification of what has been said is also encouraged by the teacher, staying with pupils' remarks to work on them further, rather than searching for or substituting another more acceptable one. Recall from the previous chapter the discussion of certain contexts in which a more explicit, context-independent expression might actually be *required*, rather than merely being a desire on the part of the teacher.

There are, however, two drawbacks to these suggestions. The first is that invoking contexts such as 'I'm closing my eyes', or blindfolding pupils, changes the activity the pupils are engaged in to one of 'let's pretend'. This can act to remove the immediacy and involvement in the situation which they are trying to gain control over for themselves and their immediate audience. Unexplained and unjustified, such constraints can seem artificial in the extreme. The second drawback is that these devices need to be employed in a genuine manner, where what happens is actually seen to be altered by pupil suggestions about the new situation. It is not much use if the announced activity is the apparently open problem of 'How to get from A to B', when the hidden agenda is a lesson on vectors which has to be reached, irrespective of the outcome of the actual discussion. An example of the difficulties of contending with activities of this sort can be seen in the next extract, taken from a report, *Four Mathematical Classrooms*, by Yates (1978).

The teacher has proposed the following situation:

> I want you to imagine there is someone sitting the other side of that door and we have got a problem. He wants to know what is on the blackboard; basically he wants to know what motorways there are, what towns there are. Can you think of a way of telling him what is on the blackboard so that he can make a copy of it?

Yates had earlier cited the example of a similar occurrence, where the same teacher had asked, 'How would you describe a cube for someone who has not seen one before?'. Both these situations seem open-ended, yet the subsequent way in which the teacher (T, here) handled the class contribution suggests that this was not the case and that he had a specific topic which he was using this gambit to gain access to and to motivate. Here is an extract from the lesson. (P refers to any pupil comment, all of which are italicized.) The *evaluation* aspect of the standard interaction sequence of Initiation–Response–Feedback, which was illustrated briefly in Chapter 2 by part of this extract, is particularly well highlighted here.

P: *Morse Code.*

T: Morse Code, well that is not necessary. We can speak to him – he is only the other side of the door. If I was to put you the other side of the door you could hear what I was saying.

P: *Coordinates.*

T: Coordinates would be one way of doing it. That would be a very good way of doing it. What do you mean by coordinates?

P: *Say five across and down this way.*

T: Well that is a very good idea, it is one I had certainly not thought of. Any other bright ideas?

P: *Hold up a mirror.*

T: Hold up a mirror . . . it cannot go through a solid door.

P: *Get one of those cameras that takes a picture in one minute and push it through the door.*

T: That is original – trust him to come up with it. Well that is another idea. Let us say that our friend has never heard of coordinates and has not got a camera. What are we going to do now?

P: *Explain it to him.*

T: Explain it to him. How are you going to explain it to him?

P: *In angles.*

T: In angles.

P: *Tell him the top angle.*

T: Tell him what top angle?

P: *The one up by Liverpool. That one, that one or that one* (points to them).

P: *All of them.*

T: Let us put it like this, whichever way we choose, apart from Terry's coordinate idea, it is going to be pretty difficult. But there are various things we can do that will help him. For example, we can say that there is a motorway from London to Bristol, we can say that there is a motorway from London to Southampton, there is not one from London to Leeds. Now all of these things are going to be important to you, are they not? At least I hope they will and there is no direct route from London to

Liverpool. You can go via Birmingham, you can go via Bristol, but you cannot go straight to them on the motorway. What do mapping diagrams means to you? Do you remember what a mapping diagram is? Look back in your books and see if you can find an example of a mapping diagram.

Yates comments, 'And so the pupils were told how to write out a direct route matrix for the network. They copied the motorway diagram from the board. They did another one similar to the first one. But it seemed to me that there had just been a lot of talk to no end.' Elsewhere, Yates remarks, 'the obvious solution was not accepted – take a photograph and push it under the door – the quickest way when a polaroid camera is available.' The purpose for this activity from the teacher's point of view was not shared by the pupils and so their responses were based on taking the activity at its face value. They also may have learnt something about the value of such 'discussion' as well as the worth of contributing to it and engaging with a question.

The I–R–F framework allows the teacher to retain control of the discourse, but also allows particular responses to be over-ruled or ignored by the teacher. All the initiation comes from the teacher, and all comments are directed to him. Deflection by some means is necessary if the teacher is to be freed from this straitjacketed form of interaction.

So far I have explored some of the gambits, verbal and non-verbal, which can be seen by examining closely video and audio records of lessons. In the next section, I want to turn to a less obvious phenomenon (one covert aspect from the chapter title), to explore the possible ramifications of a particularly common and widespread linguistic usage in mathematics involving the pronoun *we*.

3 Who is 'we'? Overt and covert messages

Mathematics classrooms, though impoverished linguistic environments in many respects, are rich in strictly-observed verbal rituals and exchanges. The title of this section is not so much an agonized cry for group identity in the face of an uncomprehending world, as a simple request for a referent. Consider the following classroom excerpt, taken from a lesson involving ten-year-old pupils, and attend particularly to the pronoun͗ employed. As you read it, think about the question: 'To what *community* is the teacher appealing when using the word *we*?' In the transcript, pupil remarks are italicized, and until the very end, the discussion is between the teacher and a single pupil. The problem under discussion is 26 − 17.

T: Can you get out your workbook please, I want to do some of those take-aways with you now. . . . Don't disturb . . . people please.
Now, we got up to these two, didn't we.

P: *Yes.*

T: Right, can you remember? OK, you ta . . . you start it off and tell me what you are doing.

P: *Put one there.*

T: No. Let's start from the very beginning. Six take away seven. Can you do it?

P: *No.*

T: No. Why can't you do it?

P. *Cos it's . . higger number on the bottom.*

T: All right. Because six is smaller than seven. All right? So what do we do then? We go to the . . .

P: *The . . . the units.*

T: No. What column's that? The tens column. Right. And what do we do there?

P: *We cross that one out . . . and then we put one there.*

T: We take a . . .

P: *Er . . . er . . . er . . .*

T: We take a . . .
What do we take from the tens column?
We take a ten, don't we. One ten. All right, take one ten from the tens column.

P: *Put one there.*

T: Yes you've got one left there. And where do you put the ten you've taken?

P: *There.*

T: You put them in the units column, right.
How many units have you got?

P: *Twenty-six.*

T: No. Put your ten in the units column. No. No. Come on,

you go to the units column and you take a ten. Where do
you put the ten?
We put it in the . . . units column, don't we.
Like we did there, and there, and there, and there.
Now how many units have you got there in the units
column now?

P: *Sixteen.*

T: Do you know where they came from?

P: *Tens, . . . and six.*

T: Yes . . . How many units in a ten?

P: *Nine.*

T: How many units in a ten?
How many units in a ten?

P: *Sixteen.*

T: No.
(to another child) Can you pass me a ten block, and ten
singles as well? . . . And ten single units please.
No, go and get me ten single units as well.

P: (Another pupil, almost inaudible) . . . *chocolate.*

T: Pardon?

P: (Same pupil) . . . *Shall I tidy it up?*

T: No. Get on with your sums. We tidy up at the end, don't
we.

This transcript contains many of the gambits mentioned or
explicitly discussed earlier in this chapter, including those of
teacher reformulation of answer ('bigger number on the bottom'
becomes 'six is smaller than seven') and 'clozed' questioning (one
word missing, to complete the teacher's sentence, which has the
effect of both controlling and shaping the allowed responses), as
well as certain ritual phrases which were repeated over and over.
My overriding impression was that of someone being coached
from a script, someone who has yet to learn their lines properly.
However, I was also forcefully struck by what seemed unusual
uses of the pronoun *we* by both teacher and pupil.

Before going on to consider some of its possible ramifications,
or how widespread this phenomenon might be, what possible
interpretations of this usage are there? Linguists recognize many
different contexts of use, where *we* is employed differently from
the customary interpretation of a plural group, including the
speaker, as referent. Wills (1977, p. 279) remarks: '*We* seems to

have the greatest imprecision of referent of all English pronouns, and therefore is the most frequently exploited for strategic ends.' It is this ambiguity concerning who *we* is, whether or not the variation is deliberately employed by the teacher in the above extract, that forms the central concern of this section.

I intend firstly to illustrate a range of contexts in which a non-standard interpretation of *we* is feasible, then to provide a description of possible intent by the speaker, before finally exploring the scope for interpretation by the hearer.

1 Baby talk

There are many systematic shifts of pronominal usage when talking to very young children. 'We're getting you dressed, aren't we' was uttered in a situation where the baby was not actively contributing at all, either physically or linguistically.

2 Hospitals and doctors' surgeries

The sentence 'Now we are going to take our temperature' provides an example of a non-standard referent. An alternative to the first sentence, with a similar meaning but different connotations, might be 'Now I am going to take your temperature.'

3 Formal discourse

'In our attempt to analyse addition and multiplication of numbers, we are thus led to the idea . . .' (Fraleigh, 1967, p. 5). Such usage can be found in books, articles or read lectures. The editorial *we* provides another instance.

4 School use

Joyce Grenfell's teacher monologues provide some delightful examples. 'Hazel, what do we do with our heads? We hold them up.' 'Susan, we never bite our friends. And Sue, we don't want GRUMBLERS in our fairy ring, do we. We only want smilers.' (Grenfell, 1977, p. 23). Other examples will be discussed later.

There are numerous other occasions when unusual pronominal usage occurs, but the four described above include some of the more familiar ones. I now intend to take the given examples and explore possible interpretations of both their intent and effect on hearers.

Baby talk involves a pervasive substitution, and perhaps arises from a desire to involve a non-languaging child in an interactive dialogue, by fusing speaker and hearer into a single whole. *We* can refer variously to the speaker, the hearer, or the speaker and the hearer together. One later aim might be that of eliciting co-operation by implicitly involving the other in the activity.

The medical example admits more than one possible interpretation. The first involves questions of power and dependence, with *we* possibly employed as a means intended to diminish the authoritarian intrusiveness of the situation, by attempting (or pretending) to make it a collaborative venture. (It is your temperature, but I am going to take it.) Other instances of this intent can be found in the softening of commands, by first structuring them with *we*, and then transforming them into apparent questions. 'Go' (imperative) → 'Let's go' → 'Let's go, shall we?'. An alternative interpretation is that the dependency inherent (and rigorously maintained) in hospital situations evokes 'baby talk' between adults, possibly in an attempt to elicit co-operation. Whatever the intent of the user, the reported effect on many hearers is one of condescension, of being talked down to and not treated as adult.

The traditions in textbook writing are so strong that, for example, the mathematician David Fowler, in the preface to his book *Introducing Real Analysis* (1973, p. 8), comments: 'I, the author, address you, the reader, in a way that may be considered unseemly by my colleagues.' Part of the effect of using *we* is to move away from the individual as mathematician in a way that I intend to explore more fully at the end of this section. Again, it seems to have to do with both power and dominance, and attempts to enrol the (at least tacit) acquiescence of the reader in what is being expressed.

The instances from Joyce Grenfell capture beautifully the classroom language of social convention, while the monologue form emphasizes the perception that her charges were virtually without language. In the classroom transcript provided at the outset, even a crude word count demonstrates the extent to which the teacher talk dominates the pupil contribution.

There is at least one instance of social convention in that

transcript. 'We tidy up at the end, don't we' seems a clear-cut example of a social convention being conveyed: that is to say, the teacher was laying down how things were to be done, in *that* classroom at least. Although it has the surface form of a question, the tone with which it was uttered makes it perfectly clear that it was not intended as such. On the contrary, it was a strong underlining of social practice. Even in such a conventional use, the scope of *we* may be restricted, in that a teacher might well say, 'We hang up our coats over there', indicating desired practice for the pupils, despite the teacher's coat perhaps having a preferential hanging place elsewhere.

What is to be made of the other occurrences of *we* in this transcript? If they represent the customary plural usage, then the teacher is incorrect to use *we*, because most of the evidence points to the pupil *not* being aware of what to do. The sense to be made of the pupil usage of *we*, other than as an instance of having learned the sort of things you are supposed (allowed?) to say in mathematics lessons, is even less clear.

'We take a ten, don't we' has exactly the same surface structure as the above social convention uses. What if pupils were interpreting all such constructions in the mathematics classroom as indications of social practice, as 'how things are to be done here'? What message about the nature of mathematics and mathematical knowledge might be conveyed? What would be some of the effects resulting from a pull-over of interpretation, namely where all non-standard *we* utterances are perceived as indicating social conventions?

One aspect of social conventions is that no explanations or justifications need be given; they just are. Such a construction can be used to convey the way things are not being done, but *should* be being done *here*, and hence can have an implicitly local feel. A pupil's experience of shifting classroom convention from lesson to lesson, both social and linguistic, testifies strongly to this.

Given a class of pupils all of whom are doing something incorrectly, a teacher can still say, 'No, what we do is . . .'. To reiterate my question posed right at the beginning of this section, who is the community to whom the teacher is appealing, in order to provide the authority for the imposition of a practice which is about to be exemplified? Are parts of elementary, or even higher, mathematical practice largely conventional, on a par with putting the date on the top right-hand corner of the page? A second major consequence of employing *we* to introduce mathematical conventions is to convey the message that there are, in sociological jargon, *in-groups* in mathematics, where 'doing it right' has solely to do with conforming to the uniform practice of this elite group. Does the authority of a mathematics teacher derive from membership of this group whose edicts are being communicated? In this light, recall Stubbs' remarks about the sociocultural function of a teacher's language.

The personal in mathematics

Much of mathematics teaching is conducted in an atmosphere of 'what the teacher does is the way everyone is supposed to do it'. Yet here is one instance where the strategic possibility of using *we*, without necessarily including the speaker, can have an interesting effect. A teacher could say, 'this is the way we do it', for instance subtraction by decomposition, without necessarily revealing how they themselves *actually* do it (perhaps by adding on). This possibility highlights differences among how I do something, how I think I do something and how I think something should be done.

This concern might go towards accounting for a common fear in mathematics of involving, and hence exposing, the self. The public image of mathematics is of something objective and absolute, permanent and impersonal. The inner mental activities of an individual are subjective, partial and relative. In the light of such beliefs about the nature of mathematics, it would therefore be reasonable to consider it not only appropriate, but the only proper way of behaving, for a teacher to refrain from exposing any personal images or thoughts. This provides another illustration of Thom's claim about mathematical pedagogy resting on a philosophy of mathematics. Acting on this belief may fail to communicate the possibility that people are working on images when doing mathematics.

This tension, between the teacher as individual and as a carrier of the general, arises in other areas of teaching. For example, a colleague who spends much of her time teaching advanced English to non-native learners, differentiates between '*I* would say . . .' and '*we* would say . . .'. The former phrase she uses to indicate something about the way *she* uses the English language, about her idiolect. The latter she uses to convey her beliefs about judgments of grammaticality which she feels would be shared by the majority of native speakers of standard English.

The same can be true of accent in foreign language teaching. Are pupils supposed to model their pronunciation exactly on that of their teacher, despite an awareness that no two people speak in an identical manner? A third instance comes from the teaching of a musical instrument, such as the piano. How is instruction to proceed, other than merely trying to copy identically what one's teacher does? It is often difficult as a learner to distinguish between the general and the specific, or even to become aware that such a distinction exists.

Thus, one use of *we* rather than *I* is perhaps intended as a clue to generality. It can also be employed as a means of spreading responsibility, while at the same time deriving weight and authority from (large) numbers. Such intents can be discerned in politics; 'We, at the ministry, believe . . .', or perhaps in speaking as a true representative on behalf of one's constituents. If a referendum had been taken, and if honest, politicians could use *we* in such circumstances, even if they personally are dead against the proposal. Unfortunately, the reverse situation is more likely, where the representative gives a personal opinion, yet passes it off as a group consensus. Does the mathematics teacher as politician seem a useful perception? Certainly the same tension between the individual and the representative of a community seems present in both contexts.

Mathematics as a spectator sport

Recall the aphorism I cited in the Preface concerning the apparent non-existence of people in mathematics. There are considerable pressures exerted by the approved style of communication for mathematics to be viewed as a depersonalized subject, but also one where there is an Authority. The widespread use of the passive mood (as in Science, where expressions such as *the crucible was weighed* rather than *I weighed the crucible* are part of the approved style) allows the active agent effecting the change to be omitted altogether. There is also the equally common use of the imperative in mathematical discourse (e.g. *consider*, *suppose* or *define*), which permits a similar massive deletion of linguistic elements. Another form is

'*Let* x be the number of . . .'. Who is giving the permission for this to be done? The mathematical imperative (with apologies to Ardrey, 1967) is a topic worthy of considerable attention.

In this section, I have attempted to illustrate how the use of *we* can be seen as an imposition; one which fails to take into account the wishes or interests of 'participating' individuals. This effect is compounded by the similar tradition in textbooks whereby the author is *we*. A famous undergraduate textbook on algebra by Fraleigh (1967) has a section entitled 'Our basic goal achieved', and his book contains many instances of interesting pronominal usage:

> In our attempt to analyse addition and multiplication of numbers, we are thus led to the idea that. . . . As mathematicians, let us attempt . . . (p. 5)

This example follows a paragraph in which *one*, *you*, *I, the author* and *mathematicians* have all been used, though I must admit to a certain confusion in places as to precisely who is being referred to at any particular instant.

> We have seen that any two groups of order three are isomorphic. We express this by saying that there is only one group of order three up to isomorphism. (p. 57)

In the first sentence, Fraleigh seems to be expressing a hope about the mathematical state of understanding of his audience, while in the second, he is providing the accepted terminology that *we* use. To end the extracts from this text, here is one which includes many of the preceding items:

> In the past, some of the author's students have had a hard time understanding and using the concept of isomorphism. We introduced it several sections before we made it precise in the hope that you would really comprehend the importance and meaning of the concept. Regarding its use, we now give an outline showing how the mathematician would proceed from the definition. (p. 56)

The effect on me of reading this book was to emphasize that choices had been made, ostensibly on my behalf, without me being involved. The least that is required is my passive acquiescence in what follows. In accepting the provided goals and methods, I am persuaded to agree to the author's attempts to absorb me into the action. Am I therefore responsible in part, for what happens?

Such linguistic conventions beg the question of the relation of the individual to the mathematics under discussion. Suppose it were widely accepted that mathematics is essentially social in nature, and communication about mathematics requires genuine negotiation and sharing of meaning. How might such a conception of mathematics be conveyed through a written medium, with the reasons for particular conventional agreements being communicated and explored? Any proposal would need to contend with the phenomenon of text authority alluded to by Thom when he claimed (1973, p. 196) that, 'As soon as one uses a textbook, one establishes a didacticism, an academicism, even if the book be so written as to promote individual research.'

A more important question is that of mathematical practice as a whole. How might it be studied? With an emphasis on communication, what are the rituals surrounding such practice, linguistic and otherwise? Do pupils have to learn to attend, read, write and speak in particular ways in mathematics classrooms? What occurs when such rituals are broken? (See Davis, 1966, and Brown, 1981, for two examples of 'ritual speech' being broached, the former involving 'eight from four, you can't'.)

Summary

This chapter has examined a number of spoken linguistic threads which emerge from the theme of classroom interaction about mathematics. I introduced the term *gambit* to describe particular teaching strategies which were explored in terms of both sacrifice and long-term advantage. Some traditional teaching practice might well benefit from also being looked at in these terms. In the context of asking questions, the so-called 'clozed' style provided a particularly clear example with some definite sacrifices, and I looked at the notion of question deflection.

Certain further aspects of classroom style of interaction were exemplified by looking at transcripts, and I ended by attempting to express some thoughts about the nature of mathematics which arose as a result of noticing a particular linguistic practice at work, namely the use of *we* in explanations. By focusing attention on the identity of this *we*, I found some unsettling questions arising about the nature of mathematical knowledge as well as social, community-orientated aspects of mathematics and practices of mathematicians.

In the next chapter I shall explore the notion of the specialist language aspects of spoken mathematics which were remarked on in the example containing the discussion of *sector* and *section*. The central idea will be a linguistic one, namely that of a *register*, which will be used to describe more precisely the position of mathematics in relation to English – one commonly, if not very accurately, referred to as that of being a 'language within a language'. It will pinpoint many of the differences in usage (such as the use of the passive and imperative forms which I have already identified here), and explore some of the difficulties which arise from this, with the notion of metaphor appearing explicitly in this context.

4

The mathematics register

There is a vague, often impenetrable, no-man's-land between the discourse of poets, philosophers or people in general, and the discourse of mathematicians. People say things multiply when there is increase. Mathematicians also say they multiply when there is decrease (times half) or when neither increase nor decrease is in question (times a matrix).

Dick Tahta, review of *Infinity and the Mind*

Most mathematics classes in Britain take place in a mixture of what has been called 'ordinary English' and 'mathematical English' (Kane *et al.*, 1974), where the latter refers to the use of the English language for mathematical purposes. Many confusions occur as a result of differing linguistic interpretations, where the teacher, for instance, might be employing terms from what has been loosely called a mathematical 'dialect', with the pupils interpreting everything they hear as ordinary English, thus trying to use non-mathematical meanings in a mathematical context.

Several levels of formality are possible in either mathematical speech or writing. It seems clear, however, that mathematics is *not* a dialect in the sense of a regional language variant such as Geordie or Cockney. A *dialect* is a variant of a language according to the user; thus individuals or groups use dialects. There is another contrasting term, namely *register*, which seems more appropriate to the situation of mathematics in relation to English. *Register* is a technical linguistic term which Halliday (1975a, p. 65) describes as 'a set of meanings that is appropriate to a particular function of language, together with the words and structures which express these meanings'. Hence, unlike a dialect, individuals may adopt a particular register on certain occasions and not on others.

Access to particular registers, and also awareness of which

75

elements and constructions belong to which, can be extremely important. An instance comes from Taylor (1984, p. 8), who complains,

> Take, for example, the way in which contemporary educational discourse has become politicized. Words such as 'child-centered', 'unstreamed' and 'creativity' on the one hand, and 'basics', 'core curriculum' and 'excellence' on the other have become the property of left and right respectively, serving to label the political and social affiliations of those who employ them.

Much more seriously, Nazi interrogators were said to have sent for torture, as suspected Communists, individuals in whose ordinary conversation a particular word appeared more than very occasionally. Its presence was taken as evidence of many conversations of a political nature. That word was *korrekt*: the closest translation I can find possessing similar political overtones is *sound*, in the sense of *ideologically sound*.

The notion of *register* has been widely employed by the linguist Halliday, although the particular linguistic features he stresses as central to the notion of register have altered with time. His initial description was in terms of variation in lexical items and grammar. However, he later emphasized variation in actual meanings, rather than solely surface variation.

> We can refer to a 'mathematics register', in the sense of the meanings that belong to the language of mathematics (the mathematical use of natural language, that is: not mathematics itself), and that a language must express if it is used for mathematical purposes. . . . We should not think of a mathematical register as consisting solely of terminology, or of the development of a register as simply a process of adding new words. (Halliday, 1975a, p. 65)

Thus it is not just the use of technical terms, which can sound like jargon to the non-speaker, but also certain phrases and even characteristic modes of arguing that constitute a register. Part of learning mathematics is learning to speak like a mathematician, that is, acquiring control over the mathematics register. I discussed briefly in the last chapter the way in which teachers take seriously the function of transmitting this register.

In this chapter, I want to explore some of the semantic and educational consequences of the existence of a mathematics register in English. Both teacher and pupil may well be aware that in school, particular deep-seated linguistic conventions and expectations pertain which regulate many of the interactions which take place. (This has frequently been termed *the language of the school*, although this too is actually a register.) Part of language competence involves knowing what is socially acceptable in such linguistic situations, as well as having access to the required variants.

Are pupils sufficiently aware of the existence of a mathematics register which is employed within the school? It is one thing to be aware that a changed style of language behaviour is required when moving from, say, a history class into a mathematics one, but a register involves more than just stylistic differences. What semantic variations, that is, what conceptual ones, are there? How widespread and pervasive are these differences on the one hand, and how insidious are they on the other, by not being readily apparent?

In particular, how do pupils keep such meanings and uses separate and how do they acquire knowledge of when to invoke one meaning rather than another? How aware are teachers themselves when they are using various registers? Do teachers explain these differences to their pupils, or do they assume that their pupils know which items are occurring in the mathematics register, and when using this particular register is appropriate? These questions, while not all specifically addressed in this chapter, are central to any understanding of the acquisition of the mathematics register by pupils, a topic of considerable importance for mathematics educators.

1 The nature of the mathematics register

The requirements of the expression of mathematical ideas in natural languages lead to the development of mathematics registers in which discourse about mathematical ideas, objects and processes can take place. For the English language, this evolution has occurred over many centuries. Once a mathematics register has developed, certain meanings will then be available in the language.

One obvious means by which a register expands is the coining of specialist terms, the invention of new words referred to by Halliday above. In that article, he describes a number of different means by which *new* terms may be created, as opposed to the reinterpretation of already-existing words. Many different disciplines employ specialist terms which do not have meanings in ordinary language, and which may well be unfamiliar to a large section of the population. Words such as *tort* and *estoppel* in law, *clef* and *breve* in music, and *byte* and *pixel* in computing provide instances. A familiar feature of the language employed in mathematics classrooms is the widespread use of specialist terms such as *quadrilateral, parallelogram, hypotenuse* and *multiplicand*, words which are unlikely to be encountered outside them.

The processes by which a register can be created are quite varied. In the case of mathematics, the register's most striking characteristic is the number of terms it contains which have been 'borrowed' from more everyday English. Examples of such words include: *face, degree, relation, power, radical, complete, integrate, legs, product, moment, mean, real, imaginary, rational* and *natural*. Mathematics is not the only subject which 'borrows' words. The legal, musical and computing registers, for instance, also contain everyday terms to which specialized meanings are ascribed. One example from each area is *bar, bar* and *memory*. However, the extent to which this happens in mathematics, coupled with the fact that it is not just certain nouns and verbs (e.g.. *a ring* or *to differentiate*) to which this borrowing applies, but that it also involves a wider range of grammatical constructions, makes a quantitative difference into a qualitative one.

Such usages change over time, illustrating the development of the mathematics register in English, as well as carrying with them a sense of how these concepts were once perceived. Fowler (1973, p. 26) points out that, 'mathematical nomenclature in general is a fascinating record of long-forgotten prejudices and arguments!' The verbal trace remains, even after the psychological support offered by the particular image is no longer appropriate, perhaps after the old name has been used in a number of novel contexts. One possible contemporary instance of a term losing its everyday referent is *clockwise*, in that, in the near future, all clocks are likely to be digital. The term, however, is so entrenched, and the notion of orientation so important, that I suspect that the word will remain.

What other characteristics distinct from everyday language usage are there of the mathematics register, apart from the reinterpretation of certain existing nouns and verbs and the coining of new terms? More problematic in many ways, partly because the alteration in meaning is less apparent, are differences regarding the meaning and use of words such as *some*, *all* or *any*. Tall (1977) has investigated first-year mathematics under-graduates' interpretations of the words *some* and *all*, and discovered that for many students, the terms *some* and *all* are contrastive rather than inclusive, i.e. *some* entails *not all*. Thus, for instance, the statement 'some rational numbers are real numbers' was regularly judged to be false, because *all* rational numbers are real numbers.

One difficulty with the word *any* can be highlighted by comparing the following two questions.

Is there any even number which is prime?
Is any even number prime?

The first question seems clear and could reasonably be uttered in despair at finding many examples of odd primes but no even ones. The second can be interpreted in two conflicting ways:

Is any (one specific) even number prime? Answer: Yes, 2 is.

Is any (i.e. every) even number prime? Answer: No, almost all are not prime.

Mathematicians tend to use *any* to mean 'every', and, on occasion, their meaning conflicts with ordinary usage. An Open University Foundation course assignment read as follows:

W stands for the set of all 2×2 matrices

$$\begin{pmatrix} \dfrac{a}{a-2} & \dfrac{1}{2-a} \\[2mm] \dfrac{2a}{a-2} & \dfrac{2}{2-a} \end{pmatrix}$$

with $a \in R$, $a \neq 2$. For any matrix A in W, show that $A^2 = A$.

Answers submitted from tutorial groups contained particular (numerical) matrices for *A* (each student had chosen a different one) for which they had derived the result in their chosen instance. When asked about this, one student commented 'Well, it said to show it for any, so I just picked one. I thought that was what it meant.' For a further exploration of this area, see Mason and Pimm (1984).

Conventional prepositional usage can also cause difficulties. There is often a widespread sliding between a polygon as a figure made up of lines and as the associated region inside with the lines for its boundary. Thus it is customary to talk of the area *of* a triangle rather than the area *inside* one, as would be appropriate for a triangular region. The word *of* is used presumably to indicate possession or attribute. This example raises interesting linguistic questions about the perceived referents of terms employed in a mathematical context.

A second instance can be seen in the statement of Pythagoras' Theorem where small changes of prepositional wording subtly reflect major differences in perception regarding the meaning and interpretation of the result. Common variations in the opening phrase include:

The square *on* the hypotenuse . . .
The square *of* the hypotenuse . . .

The former reflects a Greek geometric paradigm, where the hypotenuse is a magnitude (a length), and the square *on* it is a geometrical square. In this interpretation, the theorem states a result about the relationship between the squares themselves. The square *of* the hypotenuse (a numerical operation), on the other hand, involves seeing the hypotenuse as a number, and the theorem is then talking about relationships between numbers.

Other problematic terms include grammatical connectives and operators such as *and*, *or*, and *not*. One instance of the existence of a mathematics register influencing attitudes about everyday speech is that of the double negative. Stubbs (1983b) explores this in terms of what he calls 'the pseudo-algebraic view of language', which equates, among other things, the logical operation of negation with the English word *not*. He points out that certain languages such as French (e.g. je *n*'en sais *rien*) and Spanish (e.g. yo *no* se *nada*) contain double negative markers without them being considered illogical languages. However 'I don't know nothing' is often seen as logically deficient, rather than being a non-standard form, with the word *nothing* either merely intensifying 'I don't know', or being automatically required for a negative in some dialects, as it is in standard French or Spanish.

Similar comments about the influence of the mathematical use apply to the propositional use of the word *if* in English. The statement,

If you want to go to Australia, then you can get information from the Australian Embassy

can be criticized as incorrect under the above-mentioned pseudo-algebraic view, because its logical reversal (the contrapositive, which in standard logic is supposed to have the same propositional content as the original statement), is clearly false. Its contrapositive is: 'If you cannot get information from the Australian Embassy, then you do not want to go to Australia.'

Instances such as those mentioned above of the pseudo-algebraic view of language are often cited as criticisms of the way pupils talk or write. This could be seen as an attempt to encourage pupils to carry the particular meanings of the mathematics register into their everyday speech, under the guise of 'logical thinking'. The construction *if . . ., then . . .* in the mathematics register *always* carries with it the force of logical implication. Mathematics at all levels relies on such formal styles of argument and hence on particular meanings and uses of these terms, since nearly all mathematical propositions fall into an 'if . . ., then . . .' format.

(In Russian, there is no article marker distinguishing *a* from *the*, the definite from the indefinite. Yet this language contains a sophisticated mathematics register fully capable of distinguishing the meaning 'there exists' from 'there exists a unique'. It is very suspect, I feel, to go from observations about the marked surface features of particular dialects or languages to conclusions about the conceptual thought that can be carried by them.)

Variations in the grammatical category of borrowed terms

The use of common English words for specialized terms in mathematics can lead to shifts in their grammatical category and function. In fact, this phenomenon acts as a further indication of a specialized usage. The simplest example can be seen with the number words themselves: *one*, *two*, *three*, etc. In ordinary English, they function somewhat as adjectives (one book, two crows, three linguists, . . .) and there is number agreement between nouns and verbs. Thus, for example, there is the adage that 'one rook is a crow, but two crows are rooks.' Rather than indicate a property of crows themselves, number words refer to a property of the collection of crows under discussion. However, number words can also serve as nouns in mathematical discourse.

In some languages, number words are more clearly adjectival, in that the form of the number word itself varies with certain features of the attributive noun. In ancient Greek or Latin, the form of the number word agrees with the gender of the attributive noun. In Russian, a reverse influence obtains in that the number words require the genitive case for the following noun, and the number words *one* to *four* require the genitive singular, e.g. *three of house*, whereas those greater than four take the genitive plural, e.g. *seven of cranes*.

Imagine now a language in which different sets of number words are used according to the physical characteristics of the objects being counted. Thus, long objects could require one set of terms, while animate ones could need a second, independent set of counting words, and so on. The contrast between this situation and that described in the previous paragraph, one hypothetical and one actual (although there exist natural languages with characteristics similar to my second instance), is whether it is the grammatical category (e.g. gender) or the meaning of the attributive noun which determines the form of the number word. In my hypothetical example, it might well be impossible to be counting intransitively: one must always be counting *something*. Therefore, if a mathematical register were to emerge in this language, either one of the existing sets of number words must develop this possibility, or a new set of number words must be created, able to carry this mathematical function.

In mathematical English, number words also function as nouns, that is, as names of objects which themselves possess properties given by adjectival descriptors such as *prime* or *even*. They are also used with predicates such as *is a power of seventeen*. These two distinct uses of number words, namely adjectival and nominal, co-exist in the multiplication tables, which include items such as *four fours are sixteen*. Just as with *four chairs are black*, the first 'four' is performing an adjectival function, while the second clearly exhibits its nominal status through acquisition of the plural marker '-s'. Thus *a four* is a singular entity of which we can have any number – recall the question 'how many fours in twenty-four?'

Further examples of variations of grammatical category can be found in the fractions terminology. The expression *three sevenths* can be analysed as the noun phrase (*Adjective*) (*Plural Noun*), (or perhaps as *Noun Noun* e.g. police car), except that plural agreement in the verb does not automatically occur. Thus 'three sevenths *is* more than two sevenths' and 'three sevenths *lies* between 0 and 1' are the 'correct' expressions in the mathematics register, although it is common to hear people say 'three sevenths are more than two sevenths.' The process of abstraction whereby 'three sevenths' comes to be seen as a single object and is on occasion written as three-sevenths, and therefore requiring a singular verb, is an extremely complex one.

This difference attests to the richness of the fraction concept and the importance of being able to move from one perception of a mathematical object to another at will; in this case from $\frac{3}{7}$ viewed as a single number to $\frac{3}{7}$ seen as three lots of $\frac{1}{7}$ (i.e. three things), in addition to the perception of $\frac{3}{7}$ as the operation of 'three seventhsing' (cf. halving).

The next extract provides a clear illustration of a comparable shift in the grammatical category of a borrowed term. The following problem was presented to a thirteen-year-old pupil.

PROBLEM
Here are some polygons.

How many diagonals does each one have?
If you knew the number of sides of any polygon, could you work out the number of diagonals?

I had assumed that there would be no difficulty with the term *diagonal*, so had not explored the concept at the outset. I invited her to read out the problem, asked her if she understood it, to which the answer was 'yes', and she then computed the results of the four specific examples as follows:

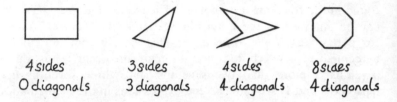

4 sides   3 sides   4 sides   8 sides
0 diagonals   3 diagonals   4 diagonals   4 diagonals

Diagram 3

My only comment at the time was a neutral 'I see', and then I directed her attention to the second part of the question (J is Jill, D is David.)

J: No, you can't, 'cause if you turn it round you get more diagonals.
D: Could you show me?
J: (Draws Fig. 1) This has four diagonals.
D: Oh, I see. Could you show me a triangle which only has one diagonal?
J: Sure (Draws Fig. 2, crosses it out and draws Fig. 3).
D: And a four-sided figure with less than four diagonals?
J: (Draws Fig. 4)
D: Yes I see. What about an eight-sided one? Could it have eight diagonals?
J: (Draws Fig. 5)
D: Could you write down for me what you have found out?
J: (Writes)

it depends on what the shape is
and which way you place it.

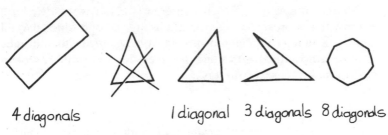

4 diagonals   1 diagonal   3 diagonals   8 diagonals

Figures 1–5

On the basis of the above I feel confident in asserting that Jill's interpretation of the term *diagonal* differs markedly from mine. Her notion seems to be something like 'sloping side of a figure relative to the natural orientation of the page'. If I had been alert to this at the time, I would have altered the orientation of the page relative to her to see whether or not it affected her usage. With this notion of *diagonal*, the posed problem has been solved completely and correctly. (Notice in passing the ambiguity of the two *its* in her written record.)

In moving from 'ordinary English' to 'mathematical English', the term *diagonal* has undergone a syntactic category shift. In the former context of use, its function is adjectival, contrasting with *vertical* and *horizontal*. Once again, in mathematics, a property of an object is used to name the object itself, and thus the term *a diagonal* (a noun) is coined. Once this property has been objectified, questions such as 'how many?' can be asked. Unfortunately in some ways, mathematical diagonals do not necessarily have to lie diagonal to the page, although many will.

When examples of figures are drawn to illustrate or merely invoke a concept, the orientation is seldom randomized, and many pupils seem to include the particular orientation in their concept. Skemp (1986) provides a nice example of a pupil having difficulty even constructing a square on the hypotenuse of a right-angled triangle in standard position, as the pupil's notion of *square* seemed to include that of the sides having to be 'square' to the page.

The referent for the mathematical sense of *diagonal* was missing from all of Jill's pictures. Pupils are taught to 'see' diagonals even when they are not drawn in and to speak of them as attributes of, for example, a square, even when they form no part of the definition of a square. This absence may account for

Jill looking for (and finding) a referent for the word *diagonal*, one which was present in the picture itself, and one which agreed with the everyday meaning of the word *diagonal*. In the absence of the mathematical meaning, the everyday meaning has been used to make sense of the task.

Further aspects of the mathematics register

So far, I have looked at two ways of expanding the mathematics register – one by extending the terminology by the coining of specialist terms, the other by reinterpreting existing terms from ordinary usage. Halliday is insistent, however, that there is much more to the development of a mathematics register than just lexical expansion.

> It is the meanings, including the styles of meaning and modes of argument, that constitute a register, rather than the words and structures as such. In order to express new meanings, it may be necessary to invent new words; but there are many different ways in which a language can add new meanings, and inventing words is only one of them. (Halliday, 1975a, p. 65)

He points out, in addition, the importance of certain whole expressions (*locutions* as he calls them), ones whose meanings cannot necessarily be understood merely by knowing the meanings of the individual words; that is, the expressions function as semantic units on their own. Mathematical examples of such composite words and expressions include such larger structural units as *if and only if*, *square root*, *absolute value*, *greatest common divisor*, *simultaneous equations*, *spherical triangle* and, more esoterically, *topological vector space*.

These expressions behave somewhat like idioms in English. The expression *he kicked the bucket* (used to mean 'he died') has a number of restrictions on it in terms of the range of linguistic transformations which can be applied while retaining its idiomatic sense. *Kick the pail* does not work (substitution of a closely-related lexical item), nor does *the bucket was kicked by him* (passive transformation). Thus although it is analysable as a composite expression, the parts in this case are somewhat inseparable. There is also the question of how much of the semantics of the expression as a whole can be obtained from the meanings of the individual component words.

In the case .of the expression *spherical triangle*, a spherical triangle is not a triangle in the shape of a sphere, although the expression *a spherical container* would mean a sphere-shaped container. Thus it is not the normal adjectival use of *spherical* which is being employed. This usage refers to a triangle *on* a sphere. The expression is still problematic, however, as now it merely seemed incoherent, because do not triangles have to have straight sides? I will temporarily postpone further discussion of this example, as examining one means by which such locutions are created will require a detailed look at the notion of metaphor, which is the topic of the last three sections of this chapter.

An occurrence of an 'innovative' locution arose from a pupil describing the following figure as a *left-angled triangle.*

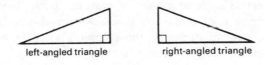

left-angled triangle   right-angled triangle

Diagram 4

Apparently unaware of the usc of the word *right* to refer to a particular type of angle, and hence its being employed adjectivally in the locution *right-angled triangle* (or its familiar contraction *a right triangle*), the pupil looked for and found an aspect of this sort of triangle in its customary orientation which would account for the appellation *right*. The created locution *left-angled triangle*, used to refer to a comparable triangle in a different orientation, accurately reflected this perception. After all, what is 'right' about the configuration is not readily apparent. The Greek origin of the term, rooted in their perception of the angle being proper or fitting, i.e. balanced and symmetric, is not an obvious one. This example also points out the general concern of how a teacher can be sure of what a pupil is attending to in a picture or diagram. What to attend to can be cued, intentionally or otherwise, by means of the particular language used.

To conclude my discussion of the nature of the mathematics register, it remains to note that the variations are not only in the meanings of the words used (and the grammatical categories fitted by such words), but also include locutions as I have mentioned above. More subtly, the deviations from 'ordinary English' also involve particular sentence constructions less commonly employed in everyday speech. Examples in mathematics include widespread use of the passive mood, gerunds such as *addend* and *integrand*, and a range of imperative forms, including *let*, *suppose*, *define* and *consider* as opening words in sentences, which I mentioned briefly in the last chapter. A study of such constructions, which point to new functions of language beyond reference, is outside the scope of this present work.

2 Register confusion

I wish to spend some time looking at certain differences in meaning between 'borrowed' elements in the mathematics register and the use of the same terms in ordinary speech. The overall effect of such large-scale alterations in meaning can give rise to a problem of *semantic contamination*, where, as we have seen in the case of *diagonal*, the common meaning or usage is employed to try to make sense of what is being heard or read. The least that can occur is the transfer of connotations, while at worst, a completely variant meaning can be employed.

Most mathematics classes are conducted in a mixture of the registers of ordinary and mathematical English, and failure to distinguish between these two can result in incongruous errors and breakdowns in communication. Below are a number of illustrations of this phenomenon, namely hearers attempting to construe what they hear in terms of the everyday meaning of some central term, or people using the same words but understanding markedly different things by them. It is only as a consequence of these anomalous responses that an awareness of the differing meanings entertained by teacher and pupil came to light.

My main point is that the processes by which pupils attempt to assign meaning to the phrases and expressions they hear in mathematics classes are completely consonant with those by which they acquired and manipulated meanings as young children 'learning how to mean'. The process seems to be as follows. To assign a plausible meaning to an unfamiliar expression or usage, bring to bear as much knowledge of the world as you have as well as knowledge of language itself. Then try it out later and experiment. That is to say, use both the context of the utterance and the syntax and semantics of the words you are familiar with as far as possible, to guide your assigning of possible meaning.

Clark and Clark (1977, p. 486) offer a couple of non-mathematical examples of this approach at work.

> A nursery school teacher divided her class into teams, spread a small blanket on the floor at one end of the room and said, 'This team will start here', and then put another blanket at the other end of the room, saying, 'This team will start here'. At home, later, a child put down a blanket and set her baby brother on it. He crawled off and the child complained to her mother: *He won't stay on my team!*

> A second example they cite was 'A mother said "We have to keep the screen door closed, honey, so the flies won't come in. Flies bring germs into the house with them." When asked what *germs* were, the child said: *Something the flies play with*.'

My first mathematical instance comes from *The Life and Letters of Lewis Carroll* (Dodgson, 1898).

> Dr Paget was conducting a school examination, and in the course of his questions he happened to ask a small child the meaning of *average*. He was utterly bewildered by the reply, 'The things that hens lay on', until the child explained that he had read in a book that hens lay *on an average* so many eggs a year.

The confusion results from a valid alternative parsing of the sentence together with the presumption of an unknown lexical item, rather than the existence of an unfamiliar expression *on average*.

My next example (Coghill, 1978) concerns an infant classroom where the teacher had been discussing with her pupils the notions of odd and even numbers. She proceeded round the room labelling the tables as odd or even according to the number of children at each. One child regularly got up and left the particular table he was at should it be about to be labelled *odd*. 'There is one boy who always has a quick "count-up" and moves to a table where he makes it into an *even* number, because he doesn't like being the "odd one".' The term *odd* can mean 'peculiar' or 'not divisible by two without a remainder', according to the context of the utterance. This pupil, who was unwilling to sit at an *odd* table, apparently had the first meaning in mind.

The teacher had explained the notion of evenness in terms of being able to share it into two groups: 'to find out if a number was an "odd" number or an "even" number, we could take that number of "things", share them between two people and if each person had the same amount it was an "even" number and if there was one left over it was "odd".' One of the exercises the pupils were asked to do was as follows. 'What is the number of your house? Is it an odd number or an even number?' A pupil replied that hers was number *15* which was even; 'we share it with the people upstairs.'

There are a number of points here. The first is a possible confusion between the mathematical and everyday meanings of the word *share*. A second is the confusion between oddness or evenness being an attribute of the house rather than of the number. For the pupil, 15 seems to be an attribute of the house, or rather, in this context, the referent for 15 seems to be the house itself, and so it is the house which is to be divided up, *shared*, rather than the number.

There are connotations of many words which can produce dissonance with the customary mathematical usage. The example of *difference* cited in the opening chapter provides one example. *Halving*, referring to a process of division into two, but not necessarily equal, pieces allows someone to reasonably speak of (and ask for!) the larger half. The expectation raised by the phrase *a fraction of its former cost* is that a fraction is both small and certainly less than one. Other connotations are invoked by the peculiarly moralistic terminology of *vulgar* and *improper*, or images such as instability engendered by the term *top-heavy*, for fractions such as $\frac{5}{3}$. We talk of *reducing* fractions to their lowest terms, yet the purpose of the transformation is to leave the value unchanged – it is an alteration in form rather than content.

A further difficulty occurs with the relation between the notions of *square* and *rectangle* (or, in a slightly different context, between *square* and *diamond*, where, on occasion, the sole distinction is one of orientation). Possibly the existence of a different word suggests different (that is, distinct) meanings (as with *some* and *all*). There is a term for the notion of a non-square rectangle, namely *oblong*, but a widespread disagreement exists among teachers, particularly at the primary level, as to whether or not this term is suitably mathematical. *Oblong* sounds like a familiar everyday word, so many feel they should not use it. The word *oblong* provides another instance where an adjective in ordinary usage (meaning 'longer than it is wide', as in *the oblong box*) has become a noun, *an oblong*, when it moves over into a mathematical context.

Another way of describing this dispute about the mathematical appropriateness of the term *oblong*, is to ask whether or not it belongs to the mathematics register in which it is believed that mathematical discussions should take place. The passage in Chapter 3 concerning the desire on the teachers' part for their pupils to speak mathematically can be seen in this new light. For one of those teachers, *sector* was a term in the mathematics register, whereas *section* was not, and part of her job was to encourage the pupils to use terms from the mathematics register.

The situation with older pupils is not necessarily any better, despite many years of exposure to and presumed reasonably successful separation of such potentially confusing terms with multiple meanings. A common mathematical practice is that of expanding the range of applicability of a particular word. One such term is *enlargement*. As with fractions larger than one, for which the term *fraction* seems inappropriate, an enlargement by a scale factor of a half looks remarkably like a contraction, yet the mathematical term applied in certain texts is still *enlargement*.

Such confusions and interference do not end even in the secondary school. Cornu (1981) examined difficulties connected with the notion of *limit* and the expression *tends to* among sixth-form school and first-year university mathematics students. He found that their intuitions and 'spontaneous models' concerning these ideas were coloured by everyday beliefs such as the notion that you cannot go past the limit, despite being able to exceed the speed limit.

All these examples confirm that multiple meaning is a distinctive problem for many mathematical terms. Particular everyday words are often chosen precisely because of some evocative image which they capture, but the match was unlikely ever to have been exact (nor was it intended to be), and the denotation alters with time. Concepts such as *multiplication* are regularly expanded in mathematics, so that even if the original context were close to the ordinary language use, it is unlikely that all of the connotations will be the same. A number of such examples will be explored in the next section, on metaphor. Even within advanced mathematics confusions can arise. There are many different meanings within the mathematics register itself for the overworked terms such as *normal*, *regular*, *similar* and *equivalent*, and the context is paramount in determining the appropriate interpretation. In topology, the terms *open* and *closed* are used as descriptors for particular types of set. However, it is not the case mathematically that if a set is not closed then it is open (and conversely), although it is very common for undergraduates to assume so. The problem of relationships which hold between terms in one context failing to transfer to a new situation will also be explored in detail in the subsequent discussion about metaphor.

One further example of a higher-level confusion is provided by the concept of *continuous*, whose English meaning is closer to the idea of *connected* in mathematics, namely being all in one piece. This is confirmed by the common use of an evocative, if not wholly accurate, paraphrase of the definition of a continuous curve as 'one that can be drawn without taking your pen off the paper'. Thus to claim that the function $f(x) = 1/x$ is continuous conflicts with the everyday notion as the graph is in two pieces. Mathematically, the definition strategy is firstly to define continuity of a function at a point, in terms of its behaviour near that point. A function is then said to be continuous if it is continuous at every point of its domain. The graph of $1/x$ is in two pieces, because the function is not defined at $x = 0$, so in particular, it cannot be continuous there. Nonetheless $f(x) = 1/x$ satisfies the definition for a function continuous on its domain.

A final, enlightening example of register confusion is afforded by a critique of certain applications of Catastrophe Theory (mainly the work of Professor E. C. Zeeman) written by Sussmann and Zahler (1978). They accuse Zeeman of precisely this sort of semantic shift for his own ends. Consider the following extract from their article (my emphasis).

> This is the discontinuity that we think about when we assert that animal aggressive behaviour is discontinuous. It is because Zeeman's words create the impression that his model explains it that some may react favourably to it [the model]. . . .

However, we must . . . be wary of first impressions derived from superficial analogies. *A function that jumps is not the same as a dog that jumps. A discontinuous function cannot be said to explain discontinuous behaviour just because it is discontinuous.* (p. 138)

This section has begun to detail instances (from mathematics at all levels) where the non-mathematical meanings of terms 'borrowed' for the mathematics register can influence mathematical understanding and frequently cause confusion. More subtle are the confusions or misconceptions provoked by altered connotations in a situation where there is some but not complete consonance between the mathematical and the everyday senses.

3 The creation of a mathematics register – the role of metaphor

In a quotation I cited earlier, Halliday alluded to various ways by which new meanings may be added to registers. Analogy and metaphor come to mind as powerful linguistic techniques for creating new meanings. They offer means by which the less familiar may be assimilated to the more familiar, by viewing the former in terms of the latter. In this section, I examine the concept of metaphor in mathematics, in relation firstly to the transfer of terms from ordinary language, and secondly to the use of metaphoric extension inside the mathematics register itself. Metaphor and analogy are figures of speech which make natural language powerful, and I suggest that there are comparable processes at work in mathematics itself, as well as metaphor being commonly employed in its teaching.

One means of coining words or expressions for a register which was not explored by Halliday in his article on the mathematics register is that of metaphor. In order to understand fully how the mathematics register in English (and, I suspect, in all languages) functions and develops, I believe that a detailed understanding of the functioning of metaphor is required. This process can be seen at work by looking at a country where a mathematics register is being consciously and deliberately developed as a result of a political decision to switch from English as a national language of education – such a country is Tanzania.

There was no word in Swahili for the concept of *diagonal*. An existing word was chosen, *ulalo*, meaning 'the longest of all'. An additional usage was developed for this word, and Swahili has increased in terms of its potential for expressing mathematical meanings. In its original context, *ulalo* refers to the rope cord which goes from one corner of a rectangular frame to the opposite one in the construction of string beds. The other cords are parallel to this central, longest one. The use of the term has been transferred out of its normal setting and into a context where there are certain similarities. However, it is not the case that mathematical diagonals are always 'the longest of all'; for example, in a non-convex kite (see Diagram 5), there are internal line segments longer than either diagonal. Thus, similar to the case of *diagonal* in English (where diagonals are not necessarily 'diagonal' to the page), there is a false carry-over of connotation.

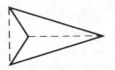

Diagonals — — — —

Diagram 5

What are some examples of metaphor from the English mathematics register? In case the above example of *ulalo* seems too unfamiliar, consider the process images embedded in the following apparently technical terms of mathematics: *tangent* (touching) and *secant* (cutting), or the graphic relationship portrayed by curves *osculating* (kissing). Access to a knowledge of Latin (the *lingua franca* of academic discourse throughout Europe for many centuries) demonstrates the metaphorical origins of these words, as well as providing sources for many words of the mathematics register in English. To someone without recourse to the Latin origins, these are merely abstruse technical terms, on a par with *pixel* or *byte* from the computing register.

Some further, more elementary examples of metaphoric usage of different kinds in mathematics include *carrying* in arithmetic, *face* in three-dimensional geometry and the use of geometric terminology in non-geometric contexts, such as referring to numbers as *triangular* or *square*. This last image can cause interference when working at the same time on configurations involving these same concept names in a geometric sense.

Consider the problem of the number of equilateral triangles needed to build the *n*th triangle in this sequence.

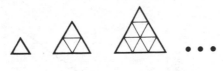

Diagram 6

Although the unit is a triangle and the geometric configurations built from such units is also a triangle, the number of triangles needed is always a *square* number (n^2) and not a triangular one.

At a deeper level, metaphor is involved in the way mathematicians discuss their objects of interest and discovery; for instance, by talking of expressions *vanishing*, functions *obeying* a rule or being *well-behaved*, the *inheritance* of mathematical properties or discovering mathematical *laws*. Mathematicians are so familiar with this way of speaking that they tend not to see it as at all problematic, or even curious.

There are two main sources of metaphor which may be of interest in mathematics education. The first consists of what I term *extra-mathematical metaphors*. These attempt to explain or interpret mathematical ideas and processes in terms of real-world events, and such metaphors can involve everyday objects and processes. Examples include, *a graph is a picture, modular arithmetic is clock arithmetic* or *a linear equation is a gear*.

The second main source, which I call *structural metaphors* and address in the next section, involves a metaphoric extension of ideas from within mathematics itself. Such metaphors are frequently also embodied in the notations of written mathematics, and examples here might include the very wide use of the '×' sign or exponential notation. (For further discussion, see Chapter 8 or Pimm, 1981.)

In either case, the metaphoric process for constructing and extending meaning is a central one. One example of an interpretative, extra-mathematical metaphor structuring the interpretations of many pupils is that of *a graph is a picture*. The widespread occurrence of such a belief, and its consequence in terms of graphical interpretation, has been well documented with children of varying ages by Kerslake. She posed the following question (1979a, pp. 71–72).

Would you tell me what you think this distance-time graph could represent?

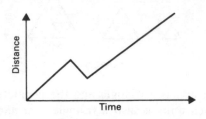

Diagram 7

Half of the children she interviewed gave responses based on this interpretative metaphor, including instances such as:

> You're going up a mountain, say, and then coming down a bit, and going up again.
> He starts at one place, goes round two corners and then on again.
> You're going N.E., then S.E. and then N.W. (sic)

Many pictograms use pictures as elements of what is being represented, for instance, one picture of a car standing for one or even ten cars. We talk of *seeing* what the function looks like by graphing it. Yet graphs are symbolic artefacts, and pupils have to learn to read and interpret them as much as other mathematical symbols. (With a velocity–time graph there is a particular problem of interpreting the *area* under the graph as the *distance* travelled, as the respective dimensions seem awry.)

Extra-mathematical metaphors may be employed personally as aids to thought and understanding, though not necessarily articulated publicly in classrooms. Papert, in the foreword to his book *Mindstorms* (1980), writes of his childhood fascination with gears, particularly differential gears.

> I believe that working with differentials did more for my mathematical development than anything I was taught in elementary school. Gears, serving as models, carried many

otherwise abstract ideas into my head. I clearly remember two examples from school math. I saw multiplication tables as gears, and my first brush with equations in two variables (e.g. $3x + 4y = 10$) immediately evoked the differential. By the time I had made a mental gear model of the relation between x and y, figuring how many teeth each gear needed, the equation had become a comfortable friend.

What moral is to be gained from this story? Papert's use of gears as a metaphor for linear equations is a personal image. Its power is rooted both in structural and relational similarities between gears and the mathematical ideas, and, crucially, in his own childhood comprehension, fascination and affection for gears themselves. Papert does not suggest that his image should necessarily be taught. One problem for many pupils could be that, although gears are a concrete and real-world phenomenon, they might prove just as opaque conceptually as the concept of linear equations. Just because gears are of the material world does not render them *inherently* more comprehensible than multiplication tables or linear equations.

My main reason for citing this example is not the particular image, but rather that his comprehension involved a metaphor which enabled Papert to develop a tool for thought. It is also important in that it provides a clear example of a child constructing personal knowledge. Would that more pupils could find functioning images of this sort which connect the ideas of mathematics with objects and processes that they feel they know and understand.

Some teachers can be reluctant to provide such unashamedly personal images, claiming that it is not at all clear that they will be usable by anyone else, and may well confuse. However, by not talking about such things at all, the existence of rich inner mental realms in which mathematics properly takes place will remain undiscovered by many pupils, who will see only the external mathematical world of symbols on paper and operations on them. Whether or not a particular offered image is successful in illuminating a concept, it at least serves the purpose of indicating that image-making is an appropriate activity for pupils to be engaged in, and that the teacher has personal images of the mathematics in question. A question requiring much further study is the following one. What are the characteristics of certain images, certain ways of 'seeing' which make them appropriate 'substitute images' in Gattegno's phrase, for widespread use in schools? An example of an inappropriate image offered as a substitute one is that of 'placing the milk bottle on the doorstep' or 'borrow one and pay it back on the doorstep' for carrying in subtraction.

Within the category of extra-mathematical metaphors there are, therefore, personal or *idiosyncratic* metaphors which perhaps one person or only a few individuals employ, and which are often invented extemporaneously, resulting from a personal insight. Papert's image of gears is such a metaphor. Another instance of an idiosyncratic metaphor is seeing the convergence or divergence of series in terms of the focusing properties of lenses (Nolder, 1984). The way of seeing embodied in the metaphor may become institutionalized (possibly by means of being written about in books!), in which case it would then become a *conventional* metaphor, a standard form both of conception and expression.

Extra-mathematical metaphors are sometimes employed overtly in classrooms. The terminology of *having* and *owing* is used in the context of positive and negative numbers. *A function is a machine*, or *an equation is a balance* provide two further very common pedagogic instances. These examples are institutionalized, that is, conventional within contemporary mathematics teaching. Some teachers may not accept that they are metaphors at all, or not consider it important that they are, but the ideas expressed by them will be familiar. Such conventional images and evocative expressions which accompany them (e.g. input and output, moving from one side to the other) are widespread.

However, it is essential for pupils ultimately to be aware that a function is not a machine, nor an equation a balance. The notions need eventually to be separated, with each possessed of some sort of independent existence. One technique for developing such an awareness is to push the metaphor to its limits to discover how a function differs from a machine. It is precisely this technique which is often most useful in order to show that the equating of function and machine, for instance, is *only* a metaphor. This emphasis on the differences may also serve to accentuate or to clarify the extent and nature of the similarities.

If the metaphor is being employed as a conceptual bridge in an introductory setting, however, such a direct comparison is inappropriate and the differences must be explored at a later stage. This is because initially the pupils will have little if any experience of functions on which to draw *apart from* that gained by means of the metaphoric image. All they may have to go on is the metaphor, and strong concepts of both are required before they can be successfully separated. The need for image assistance probably means that pushing towards the breakdown of this imagery may well prove disastrous. However, even pupils who are aware that these are metaphors and not factual descriptions, are often unclear about the respects in which the metaphors fail. For further detailed examination of this approach to mathematics teaching, see Nolder (1984); a school example provided by her is discussed in the next section.

If pupils were encouraged to formulate and verbalize their own metaphoric images then the uses and limitations of metaphor with regard to explanation in the mathematics classroom could be made much more conscious and explicit. Such an activity might be met with protests from pupils on the grounds that 'this is a mathematics lesson and not an English one', although such demarcation disputes can be useful in providing teachers with an opportunity to discover and work on their pupils' conceptions of mathematics. By the teacher inviting and discussing images, these are thereby both encouraged and legitimized. Yet one final question remains: how can you best built up an image of something you do not yet understand?

4 Structural metaphors

All the examples discussed so far have been examples of metaphoric thoughts embodied in certain expressions. In certain instances, the very language of mathematical discourse is metaphorically structured. There is a second class of metaphors, which are potentially more difficult to perceive because they rely on underlying analogies within mathematics itself. They have to do with the extension of meaning inside the mathematics register, on the basis of similarities in structure of certain concepts in mathematical theories, which therefore 'permit' the same term to be used. The *slope* of a line involves an extra-mathematical metaphor. To talk of the *slope* of a curve at a point also involves a structural one. A second, high-level example is provided by the usage of the *order* of a group and the *order* of an element within a group. Before turning to look in more detail at some mathematical instances, however, in particular that of the expression *spherical triangle*, I wish to discuss a particular view of analogy, and how metaphor relates to it.

In order to gain insight into the term *analogy*, defined in the Oxford English Dictionary as 'a resemblance of relations', it is useful to return to the Greek word *analogia*. Szabo (1978) informs us that

> It is less well known that this same word (*analogia*) was not originally a grammatical or linguistic term, but a mathematical one . . . The Greek grammarians of Hellenistic times undoubtedly borrowed their term *analogia* from the language of mathematics. So in the last analysis, we are indebted to Greek mathematics for our word *analogy*.

Interestingly, this seems to be an instance of a reverse borrowing compared with the examples discussed earlier in this chapter, namely a term moving from the mathematics register into a wider, more general use, by means of metaphor.

The mathematical meaning of the Greek word *analogia* is that of 'proportion', a relation which either holds or does not hold between four quantities. If a proportion is seen as an equality of ratios, then Szabo's observation can be expressed as follows: the relation between analogy and proportion is itself one of analogy. Thus an analogy links one relationship A:B to another C:D. For example, old age (A) is to one's life (B) as the season of winter (C) is to the year (D), or Descartes' image of being lodged in one's body as a pilot in a vessel. While proportion is a symmetric mathematical relation, the use of analogy customarily presumes a preferred direction of application, in that it assumes more is known about one relationship than about the other. By constructing this link, we can thereby better illuminate or evaluate the first relationship.

What is the movement from an analogy to a metaphor? One way of seeing metaphor is as a condensed analogy. From the analogy codified as A is to B as C is to D, we speak, for instance, of 'the C of B' or 'A is C'. Thus, 'he is in the winter of his life.' If we are presented only with a metaphor, then in order to understand it we must decide on the latent analogy upon which it is based. Our interpretation of the metaphoric expression is dependent on this reconstruction of the underlying analogy, as more than one possibility is frequently available.

Recall the example 'George is a lion' mentioned in the opening chapter. Not everything which is true of lions will be transferable, nor was that intended by the utterer. The context of use will usually allow the hearer to ascertain in the main which features were intended to be carried over and which not, though this may not always be the case. Use of the nickname *Beethoven* can be appropriately (if rather maliciously) applied to someone who is deaf as well as to someone showing a profound musical talent. Often the peculiar force of a metaphoric statement comes from the strength of the claim, namely that George *is* a lion, rather than the simile *George is like a lion*.

Spherical triangles revisited

Is *any* juxtaposition of terms meaningful? Instances abound in mathematics where apparently contradictory terms are commonly juxtaposed. The phrase *spherical triangle* provides one such expression. *Spherical*, as I indicated earlier in this chapter, appears grammatically to be just another descriptive adjective, much as *equilateral* is in the comparable expression *equilateral triangle*. Yet in the former case, the particular word applied appears to deny one of the criteria of triangularity, in that the customary definition requires triangles to have straight sides.

I wish to claim that the locution *spherical triangle* is a metaphorical condensation based on an analogy. From the above analysis it becomes clear that to make sense of this expression, we need to reconstruct the analogy on which it is based. In order to make sense of the expression *spherical triangle*, an extended sense for triangle is required, whereby *geodesic* (a path of shortest distance) replaces *straight line* (which is a particular instance of the notion of *geodesic* in the case of a plane), and hence can be sensibly applied to the surface of a sphere. There are no straight lines on a sphere, but there are paths of shortest distance, namely great circles. The underlying analogy is then

great circles : the sphere :: straight lines : the plane.

The notion of triangle previously required straight-line segments

as sides. The notion of triangle can be extended metaphorically
by means of this analogy.

The new situation can be schematized as:

triangle (defined in terms of geodesic,
applicable to any surface
where the notion 'path of shortest
distance' makes sense.)

/   \

(straight-sided)   (arc-of-great-circle-sided)
triangle   triangle

The question 'what is a triangle?' therefore has a number of
different answers, and only in the above broader context is the
locution *spherical triangle* interpretable. The expression *spherical
triangle*, therefore, incorporates a structural metaphor of considerable complexity. In the original setting, namely the plane,
which for many pupils is the only context for geometry, this
expression is a *catachresis* (the false attribution of one word to
another – see p. 106), as much as the phrase *four-sided triangle*
would be. The above example embodies a process of concept-
generalization which is a mathematical commonplace, yet its
action, function, complexity and consequences in terms of
understanding are seldom referred to or even explicitly acknow-
ledged.

What is the force of calling *spherical triangles* triangles? Such
geometric configurations of curves ('lines') on spheres may not
have had a descriptive term previously. Classifying them as
triangles results in stressing the function of this configuration
(that is, three segments of great circles meeting pairwise in three
points) in the study of the geometry of the sphere, likening it to
the role played by the concept of triangle in the plane.
Immediately whole theories, comprising definitions, concepts and
theorems line up for examination, 'translation' and exploration.
This is one primary reason for the use of metaphor, so we can
translate and extrapolate knowledge and insight from a more
familiar setting – in this case, the plane. In the process, the
adjective *spherical* becomes as appropriate or acceptable as is
equilateral as a classifier for a type of triangle.

For instance, it makes sense to talk of *congruent* spherical triangles, although there are some differences in their theory from that of congruent triangles in the plane. One example of such a difference is that the notions of similar and equilateral triangles coincide on the sphere: three feasible, specified angles can belong to only one particular size of triangle (so AAA is a congruence theorem on the sphere). A connection between the planar and the spherical context, one which may not have previously been noticed, is that a straight line is where a cutting plane meets the plane, while a great circle (a 'spherical line') is where a cutting plane passing through the centre of the sphere meets the surface of the sphere itself.

A major additional gain from this extended perspective is that the comparison also throws light on the original planar situation, and affords a perspective on the *significance* of the definitions and theorems there, which was otherwise unavailable. The significance of results in mathematics is a notion which is rarely attended to in print. One of the very few discussions of this crucial idea can be found in Brown (1978).

The example of spherical triangle highlights clearly the fuzzy edges that exist with use of analogy and metaphor. Hesse (1966) refers to those aspects of the situation which are not the basis for the initial analogy being drawn in terms of *negative* analogy (those aspects which do *not* transfer) and *neutral* analogy (those aspects which are open and undecided). The neutral analogy in the above example includes theorems about planar triangles, and they can be used to build potential propositions about spherical ones, which can then be explored regarding their appropriateness to the new setting. If true, they become new theorems (which students could find for themselves).

In summary, I hope that the above discussion of the locution *spherical triangle* has illustrated the force of Winner's claim (1979, p. 472) that

> In order to make possible a distinction between metaphorical and literal language, the definition of metaphor must be pushed one step further. It is not only the extension of a word to a novel referent on the ground of similarity that constitutes a metaphor: it is also the deliberate extension of this word ACROSS previously established category boundaries.

It is only by being consciously aware of how the meanings of the terms in mathematics change and widen under the mathematical pressures for generalization of and insight into the concepts and situations themselves, that some expressions commonly in use in the mathematics register can be understood.

The term *spherical triangle* provided one such example. A more direct instance of a structural metaphor, one originating in a school setting, has been provided by Nolder (1985), who codifies it as *a complex number is a vector*. She cites a discussion which took place between herself and one of her sixth-form pupils in a lesson in which she had been actively encouraging the class to explore methods by which arithmetical operations on complex numbers might be defined. The transcript starts with her (R) asking Ian (I) a question.

R: Any idea how you might represent a complex number using some sort of diagram?

I: I don't know if this is right, but could you draw it as a vector?

R: How exactly?

I: Well, take the real part to go across and the imaginary part to go up.

R: What made you think of that?

I: Er . . . when I started to add them I saw that diagram, you know, the one with the parallelogram. Adding them gave you the diagonal. I suppose it was the two bits, adding the real bits and then the imaginary.

R: So if you did use a vector to represent a complex number, do you think it would help you to say anything more about complex numbers?

I: Size – like the modulus?

R: Anything else?

I: Some kind of direction?

. . .

In subsequent lessons in which we continued our discussion of how a complex number might be regarded as a vector, it was interesting to note that several pupils used the term *resultant* for the sum of two complex numbers. Since the term *modulus* had been transferred from vectors to complex numbers, it was a natural assumption on the students' part that other terminology was also transferable. In such situations, it is important for the teacher to let the students see that she understands and appreciates the logic behind the use of the word and that it is not 'wrong' but unconventional.

The carry-over of terms to the new context is particularly interesting as it is precisely how extension of the meaning of terminology by means of a structural metaphor (which is so widespread in mathematics) actually happens. The remarks which I made about the subsequent separation of the conjoined items in an extra-mathematical metaphor apply equally well here.

It can be helpful to think of a complex number as a two-component vector in the plane (and the set of all complex numbers is customarily referred to as the complex *plane*), but it can also distort. For the addition and subtraction of complex numbers, this two-dimensional representation can be insightful, while for multiplication and division it can prove less so. For instance, it is possible to end up at any complex number from any non-zero complex number by means of complex multiplication (itself providing an example of extension of terminology). This fact stresses the *one*-dimensional nature of the complex numbers, and suggests that the complex *line* would be a more appropriate geometric image to be enshrined in the name. Thus, seeing a complex number as a planar vector is an extremely useful perception, but ultimately the two concepts need to be separated, with each possessed of an independent existence.

Both of these mathematical examples, namely *spherical triangle* and *a complex number is a vector*, provide instances of Lehrer's work on semantic fields (1974). She has made a study of the structure of various parts of the lexicon (e.g. colour words, cooking terms, wine terminology), and defines a semantic field as 'a group of words closely related in meaning, often subsumed under a general term'. For certain highly organized and precise parts of the vocabulary of a language (for instance, kinship terms or military ranks), this can be a useful and well-defined term of analysis, though perhaps it is less so for a language as a whole. Mathematical terminology seems to fit the requisite constraints rather well. One mathematical example of a semantic field might be the terms used for describing types of quadrilaterals within the larger field of polygonal terminology.

As well as comparing such fields across languages, Lehrer has looked at the metaphoric use of single terms and then explored how this affects the related items in the field. She has ascertained that if one word, whether as a result of metaphor or some other process, moves over to a new context, then the whole semantic field, *together with its structuring*, tends to be applied as well. In other words, certain relations that exist among sets of related terms in the original or literal context are likely to be carried over to the novel context of use.

The above example involving complex numbers exemplifies this process very well. My main point is that a mathematical area or theory provides a very clear example of a tightly-structured semantic field, where the theorems themselves provide precise details of the structural links between concepts. The carry-over of the terminology of a domain (e.g. the theory of triangles in the plane) to a new setting (by ·use of the expression *spherical triangle*, for example), carries with it the expectation that the same (or at least similar) relationships are going to hold.

One problem is that some images or relations which are pulled over may be inappropriate to the expanded or new situation. Mathematicians are sophisticated users of such a powerful technique for guiding expectation and suggesting both where to look and what results to expect. Novices less actively engaged in the creative process, or merely hearing these expressions used, can only interpret them in terms of their own experience and knowledge.

5 Some concluding thoughts on metaphor

Discussions of metaphor include the important idea of *catachresis*, namely the *false* ascription of one term to another. For instance, the relevant entry in the Oxford English Dictionary cites, 'A healthy book? Did I hear you aright?'. However, one person's metaphor may be another's catachresis, or at least incomprehension of the sense in which it is meant. Skemp (1986) provides a nice example of a child whose notion of *foreigner* requires restructuring before he can make sense of certain experiences (e.g. *himself* being referred to as a foreigner when travelling abroad). In exactly the same way, the notion of *triangle* needs to be restructured before sense can be made of the expression *spherical triangle*.

The locution *negative area* is one frequently heard in the calculus context. It arises from the claim that an integral computes the area under a curve. Definite integrals compute a number which is sometimes positive and sometimes negative and can, *on occasion*, be interpreted as the numerical value of an area. (One instance would be when the integrand is non-negative over the interval of integration.) For me, area is an intrinsically unsigned notion, and its measure should also be unsigned. This metaphor, *integrals are areas*, needs closer examination as, unexplored, it leads to the expression *negative area*, which for me is a genuine catachresis.

Examples of metaphors in mathematics often hide behind the phrase *extension of meaning*. Yet the required mental shifts involved can be extreme, and are often accompanied by great distress, particularly if pupils are unaware that the difficulties they are experiencing are not an inherent problem with the idea itself. A television reporter remarked 'The chances of finding them are, like the temperatures here, below zero.' (Bell, 1982). This might provoke the cry 'How can anything be less than nothing?'. One instance of such an extension of meaning can be found with the introduction of negative numbers, another with that of complex numbers. Calling these constructed objects *numbers* involves a metaphoric broadening of the notion of number itself. However, among mathematicians, the novelty becomes lost with time, and with it the metaphoric content of the original insight of useful extension. It becomes a commonplace remark – the literal meaning.

Why is the existence of structural metaphors a problem for mathematics education? The issue, not surprisingly, is one of meaning. Problems arise from taking structural metaphors literally, which differ from those resulting from taking extra-mathematical ones literally. If the metaphoric quality of certain conceptual extensions in mathematics is not made abundantly clear to pupils, then specific meanings and observations about the original setting, whether intuitive or consciously formulated, will be carried over to the new setting where they are often inappropriate or incorrect.

The identification, which forms the basis of the metaphor, guarantees only that certain structural properties are preserved in the extended system, while not necessarily preserving the meaning. Consider the observation that subtraction produces an answer less than the starting number. This observation arises from the metaphor identifying the mathematical operation of *subtraction* with the real-world process of *taking away*. What about $4 - (-2) = 6$? Multiplying two whole numbers together results in a number larger than the original, which is a correct concomitant of viewing multiplication as repeated addition, but this no longer holds with fractions. Why is the computation $\frac{2}{3} \times \frac{4}{5}$ still called *multiplication* and the same symbol '×' used? (In passing, note that on occasion these are distinguished at the *spoken* level, where the symbol '×' is read as *of* in the case of fractional multiplication, and *times* in the case of whole number multiplication.)

Recall the extract given in Chapter 1 concerning multiplication making bigger. That pupil had a defence against examples which defeated his expectations, namely to isolate them in a class of 'funny numbers'. Not every pupil is so resilient. The extension of concepts in this fashion can result in the destruction of meaning, if no distinction is made between the literal and the metaphorical use. Confusion arising from seeing such (often unanalysed) truths fail contaminates not only the extended system, but also the one from which it grew. The old concept resists expansion, precisely because it reflects the original and hence, perhaps, the most basic meaning.

Metaphors deny distinctions between things. Problems often arise from taking structural metaphors too literally. Because unexamined metaphors lead us to assume the identity of unidentical things, conflicts can arise which can be resolved only by understanding the metaphor. This in turn requires its recognition as such, and entails reconstructing the analogy on which it is based. Teachers will often cease to use terms metaphorically, or to be conscious of the distinction when their concept is an expanded one, but this situation will seldom be reflected in their pupils' minds.

Summary

This chapter contains some of my central ideas on linguistic aspects of mathematics. It started with the introduction of the notion of the mathematics register as one attempt to describe more precisely the relation between mathematics and English. Registers have to do with the social usage of particular words and expressions, ways of talking but also ways of meaning. Recall Stubbs' remarks which I cited in Chapter 3 on the sociocultural function of a teacher's language and the potential use of registers as territorial and status markers rather than as essential to the accurate or concise expression of mathematical ideas.

I then presented an analysis of some examples of words and expressions whose interpretation and use may well give rise to confusion. The second section attempted to delineate a wide range of problems and confusions arising from conflicting uses of the same terms. In order to obviate some of these comprehensible failures to communicate, pupils at all levels must become aware that there are different registers and that the *grammar*, the *meanings* and the *uses* of the same terms and expressions all vary within them and across them. Without this awareness, little sense can be made of a non-trivial part of mathematical usage. In my opinion, a major question yet to be explored by the mathematics education community is *how* this awareness may be fostered or cultivated.

The main topic of the chapter concerned the role played by metaphors (both *extra-mathematical* ones and *structural* ones, where the latter use ideas, processes and expressions from within mathematics itself) in extending the scope of terms used in mathematics. Metaphor, I believe, is central to the development of the mathematics register, and an understanding of the processes involved is essential to anyone attempting to make sense of mathematical speech or writing. Some of the metaphoric expressions turned out to include not merely simple nouns or verbs, but also more complex locutions such as *spherical triangle*. Some considerable analysis and 'unpacking' of the notions involved was required in order to be able to see the use of *spherical* in an adjectival sense, namely as a particular type of triangle.

The main source of the threat to understanding the basis of this aspect of mathematical language is the very elusiveness of metaphor. The same words are used throughout. If pupils are not accustomed to mathematics making sense, then having tried and failed to make literal sense of a statement, they are likely to give up, as there is nothing to indicate they are grappling with something new. This experience may even undermine their faith in the sense of situations where the *original* meaning is the most sensible one.

In conclusion, it is important to be aware of metaphors, because unexamined ones can lead us to assume the identity of elements and processes which will conflict with our past experience. Such is the power of language that almost any collocation of terms can be interpreted to some degree. However, the greatest danger is that the unexplained extension of concepts can too often result in the *destruction* rather than the expansion of meaning.

5

Pupils' written mathematical records

On parle dans sa propre langue, on écrit en langue étrangère.

Jean-Paul Sartre, *Paroles*

With this chapter, I start to look at the second main theme of this book, namely aspects of written mathematical language. What are some of the differences between spoken and written language alluded to in the Preface, and how do they affect written mathematics? Speech frequently fulfils a more direct communicative function, but because speech is invisible, it does not persist except as memories of what someone has said. Its intangible quality renders speech less permanent, and hence more readily altered and revised.

Most obviously, one key attribute of written language is that it is *visible*, and to a certain extent both permanent and repeatedly accessible. The text is complete, a static array of words, which allows the reader to step out of the linear flow. Written text is more uniform and disembodied, in the sense that it is often required to be interpretable independent of context to a large extent, certainly without the presence of the writer. Written prose tends to be impersonal, and it is harder for the author to negotiate its meaning or retain control over it. There may be no opportunity to qualify what has been written, or to react to the specific difficulties of the reader. Consequently, more commitment to a particular formulation is required, if there is to be no opportunity for later modification or explanation. It is not clear, however, that it is reasonable to compare differences between spoken and written language directly. Spoken and written texts are rarely interchangeable, in that it is seldom solely a matter of choice which determines the channel of communication. Many other dimensions are frequently present which affect the means of realization of the language, including those of private/public, informal/formal, interactive/one-way, face-to-face/disembodied,

111

spontaneous/planned and context-bound/context-free. For further exploration of these differences, as well as that of structural organization, see Rubin (1980) or Hudson (1984).

One non-interactive spoken example is provided by a telephone-answering machine. Many people report feeling very self-conscious when 'communicating' with such a device, with the result that messages tend to resemble written text in their structural aspects. One example of an interactive, face-to-face, written language context is provided by the computer. Here, the formal requirements of successful communication with the machine override many of the features which are customarily employed in face-to-face situations.

Stubbs (1980) discusses some of the many differences between the spoken and written realizations of language in general, exploring among other things the additional intellectual and social functions that a language can undertake when it has a writing system. In a summary chapter at the end of his book, he compiles a list of nine topic areas for investigation which concern the combination of literacy and classroom practice. I have listed four of them below, together with some remarks by means of which I have tried to indicate their particular relevance to mathematics teaching in schools. In the context of mathematics, however, there are also some extra factors operating which will be discussed later in this chapter.

1) *Types and uses of written material*, including texts, blackboard writing, diagrams and graphs.

2) *Pupils' reading strategies*, including notions of readability level and conventional style. Are pupils taught *how* to read mathematics?

3) *Styles of classroom language*, including the formal/informal dimension already mentioned, as well as the use of technical terminology not reflected in pupil speech. What of pupil written work?

4) *Functions of pupil writing*, including questions of purpose, audience and expected form.

There is an enormous emphasis on written work in schools at the secondary level. Not only are pupils required to write regularly, but they are also judged, often publicly, on the purported quality of what they write. As pupils move from subject lesson to subject lesson, the styles, conventions and purposes of the writing alter, while the teachers of particular subjects are permitted to remain consistently within one framework all day. Are pupils sufficiently aware of the choice of different functions and purposes which written language can serve? In the opening chapter, a perception of mathematics as being essentially written was alluded to. What reasons and pressures are there in schools encouraging pupils toward a written expression? What is the point of recording? It is to this theme that I turn in the first section.

1 Writing for myself and others

In Chapter 2 I described some of the difficulties experienced by pupils in the transition from perceiving something about a situation to expressing what has been seen precisely in spoken words. A number of reasons were put forward for saying aloud what has been seen as being of considerable benefit to the speaker. There can be comparable, if not greater, difficulties in moving from a spoken to a written representation. But are there comparable benefits?

There are many different reasons why a written record of some form may be of benefit. (The next section contains a description of various available styles.) Some of the reasons are general, some are specific to the particular constraints of the school setting, or even mathematics classes in particular; some are of value to the teacher and some to the pupil. However, written records are by no means always advantageous, and they often involve considerably greater effort, on the part of both writer and reader, than would a corresponding verbal communication. Before looking at the issue from the pupil's perspective, what reasons might teachers have of value to themselves for encouraging their pupils to record?

Discussions with groups of teachers elicited a number of reasons, including several social pressures. One of these was expressed in terms of a parental desire to see written material as tangible evidence of effort, on the parts of both their children and the teachers themselves. A second pressure is that of social control in the classroom itself. Written work makes for a silent, calm period which permits easy monitoring and is frequently justified by recourse to the prevailing view of mathematics as being an individual activity best carried out silently.

Written work is available for marking and this can be carried out in private, with reflection. The teacher is not bound up with a particular individual, nor distracted by other occurrences in the class, as could be the case with oral assessment. Written work also affords direct communication from all members of the class at once, while it would be extremely rare for a teacher to be able to talk even briefly with every pupil in the course of a lesson. It provides a certain access to how their pupils think. Pupils' errors can be informative, being illustrative of certain misconceptions, patterns of thought and beliefs. Written recording in mathematics classes can help the teacher by indicating a pupil's variant conception of some notion. Finally, written work can also provide a teacher with a visible sense of personal success.

Often pupils are unaware of the range of possible uses to which written material can be put, and this topic is not one commonly discussed in mathematics classes. What bases for response are available to teachers when pupils query why they should write something down? In the brief extract below, a teacher responded to a question with just such an on-the-spot justification. It is taken from the same lesson on pie charts as an excerpt which was discussed in Chapter 3.

T: Have I lost anyone? No, that's good. Put the heading 'example'. We'll write 'Mrs Brown has £36 a week to spend on housekeeping. This is how she spends it . . .'
P: Do we have to write this?
T: Yes. You always have to have an example, so you can look back and see how to do it.

The context was one of teacher dictation, the pupils copying the remarks written on the board into their books. The justification given for requiring such an activity in the case of an example was in terms of the reaccessibility of the information in the future. (There is another possible interpretation based on differing referents for the word *we*. The teacher's use could have meant 'I' or 'all of us', and the pupil's question could have been one of clarifying which of these possibilities was intended. If so, the second *we* meant 'the pupils'. The teacher, however, interpreted the question as a request for a justification of the activity itself.)

With written language, at least as much as with speech, there is the distinction between writing for oneself and for others. (Some authors use the term *recording* to entail 'for myself', but I have not been that systematic here.) There are a number of reasons why making written records of various sorts can be of benefit to the pupils themselves. One useful function of written records is as an external and less transient memory, where the writing serves as a record/reminder of thought. Similarities and overall structure may be easier to perceive when written down for visual comparison, and many mathematical operations (for instance, much symbol manipulation) are too complex for most people to carry out successfully in their heads.

Writing also externalizes thinking even more than speech by demanding a more accurate expression of ideas. By writing something down, it then becomes outside oneself and can be more easily looked at and reflected upon, with all the benefits of a visible, 'permanent' record. However, such writing may not be something anyone else could understand – it does not have to be context-free. This was not the reason why the marks on paper are made. They are personal, perhaps for the writer's benefit alone. There is no need for concern about the various conventions of public writing if the writing is not intended for others. If too idiosyncratic, however, it may not be reinterpretable even by the writer at a later date.

Recall the extract presented in Chapter 2 which involved a discussion of the game Leapfrog. The interchange continued as follows, containing these written attempts at expressing the generalization that they had been formulating.

T: Do you think you could write that down for me?
S: Yes sir . . . I can't write it.
T: Just write down what you said. 'Sir's got a number in his head . . .'
S: (writes) ———— × ————– = x
K: (writes) (number) × (number + 2) = number of moves
C: (writes) (———— + 2) × ———— =
T: Caroline, what comes after the equals sign?
C: I don't know.

The first thing that struck me about this extract was Stephen's apparent realization of his inability to write down what he had just said. There are at least two possible interpretations of his remark. The first is that there was a lack of stability in his articulations which meant that he was unable to reproduce what he had just said, even orally. The second is that, despite the cued request from the teacher, 'Sir's got a number in his head', that a verbal description on paper was what he was being asked for, Stephen seems aware that written representations in mathematics are not just words written down. He could have been expressing the fact that he did not know how to represent his generalization in written symbolic form.

The versions he and his fellow pupils came up with suggest this second interpretation, in that they all attempted some concise formula as their written account of the pattern. Each attempt involved some of the symbolic machinery of written mathematics. No reason had been provided as to why they should make such a record, or who the audience for the written message might be, in order to guide their decisions as to form, style or degree of completeness.

This is one instance where, just as with the question of for whose benefit spoken remarks were being made, the question of for whom the pupil is writing comes into play. I suggest that it is often unclear to pupils for whom (audience) or why (purpose) they are recording. Yet any question of adequacy, upon which judgment in class is regularly passed, is crucially dependent on both of these factors. Lacking clarification on these two basic components of any attempt at communication, teacher comments such as 'this is not clear enough' may make little sense.

What is the teacher's purpose here? When writing things down, many elements can be omitted. Pupils unaccustomed to or unconvinced by the need for written communication to stand on its own will frequently continue to operate as if in a spoken mode where words, pictures, emphasis and gesture all blend in an attempt to convey the desired message. In an oral class situation, the teacher may well provide a sympathetic audience, encouraging and interpreting what the pupil might be trying to say. With written work, however, the teacher is frequently on the lookout for potential ambiguities, and a completely different set of criteria are in force.

In school, pupils are often writing for someone else's benefit rather than their own, and the reasons for this frequently have to do with grading and testing. Rough work and jottings have personal significance, but are not intended for a wider audience. Yet once something is written, it can be thought of as 'out there', in the public domain, and somewhat divorced from the writer's control. There is a need for pupils to acquire pragmatic judgment both about the purposes of written work even more than with speech, and the criteria regarding acceptability. A sense of audience is also required which is often slow to be acquired. Re-reading from such a distanced standpoint (to see if others will be able to make sense of what has been written) is essential.

Despite earlier comments about the value of recording as a means of externalizing thought, when caught up in a rush of ideas, having to record can act as a serious brake. It can also prove very hard to re-read what you have written. One reason for this may be the difficulty of reading *locally* what is *globally* understood; that is, it is hard to stand back and re-read what you have written from the neutral standpoint of an uninformed reader. In part, re-reading can also force recognition of the inadequacy of the product – ideas which seem clear and inter-connected in the mind, when expressed on paper seem isolated and poor shadows of their former selves. This feeling of inadequacy can act strongly as an anti-checking force.

Models for public writing

What of the *teacher's* writing of mathematics? Teacher writing is most often perpetrated on the board or on handouts, and hence is writing for public consumption and, often, for copying and emulation. Therefore there is an expectation that it should be perfect. It is quite a common practice for teachers to request monitoring of board writing for mistakes of any sort, in part to encourage attention and detailed concentration. Similar remarks hold for textbooks, where the exposition is planned, polished and coherent, intended to be understandable at first reading.

Unfortunately, this same public aesthetic can carry over to pupils' written work, which seldom carries the same transmission functions. A strong temptation is for pupils to want their own writing to be seamless and fault-free – hardly surprising since the only comparisons they can make are with text or teacher writing. This is particularly evident in mathematics, where there are still traditions of neatness, legibility and no mistakes or crossings out as common criteria for acceptable written work.

These precepts can be traced back at least as far as Thorndike's (1922) injunction never to show a wrong method or let pass a wrong answer for fear it may reinforce something incorrect. The avalanche of 'snopake' in pupils' exercise books is the most recent embodiment of this tradition (though also reflecting an adolescent preoccupation with neatness). The evidence is destroyed in a Watergate-style obliteration of anything which might serve as a process clue.

Although there are numerous benefits from writing things down, it should be borne in mind that considerable mental effort is involved in writing. It is possible to lose sight of what was trying to be expressed because the problems of writing itself take up all of conscious attention. Expression through the written medium can be both time-consuming and arduous. Many adults and children alike are reluctant to write things down, particularly in situations where there is no clear reason why they should. To repeat my earlier question, to what extent are pupils aware of the benefits, and do they take them into consideration when asked to write something down? In the next section, I examine the different forms of mathematical records that pupils make in response to such a request to write down what they have found out.

2 Styles of recording

This section attempts to catalogue and classify the range of recording styles which may be observed in pupils' written mathematics. Although there is in fact a continuum of styles, I shall exemplify and name three positions on it, using the terms *verbal*, *mixed* and *symbolic*. The symbolic provides a high-status, written recording style towards which most mathematics teachers are aiming. Unfortunately, many attempt to move pupils along the recording continuum toward the symbolic end much too rapidly.

From the outset, I wish to stress that *all* the recording styles which I shall discuss in this section have the *potential* for accurate and precise written records of mathematics (an important topic explored further in the next section). The singular advantages of the symbolic form may overcome the clarity of expression of the fuller, more verbal forms only when the expressions have to be transformed.

The first style I call *verbal*, and it involves no special mathematical symbols or signs, the whole written account both of problem and solution being solely in recognizable words together (perhaps) with symbols for numbers. Here are four illustrations of what I mean by a verbal recording style. An eleven-year-old pupil had been asked to write down how to find the area of a square for another pupil who did not know. (This technique of asking pupils to act as informants for others is quite commonly employed to provide some contextual background for requests of this type.) She wrote:

> To find the area of a square you multiply the length by itself.

A board in the back of a Coventry taxi declared:

> Fares outside the city limits will be charged at meter rate plus one third.

The Cockcroft report cites a clerk describing what calculations her work involved.

> To get the rate of pay per hour, we add together the gross pay, the employer's National Insurance and holiday stamp money. Subtract any bonus and then divide by the hours worked.

In this last instance, when spoken, there is no guide to the internal representation from which the clerk is working. When written in the above form, the style is completely verbal, every word is an English one and involves no special symbols. As a final example of a verbal style, a twelve-year-old pupil produced the following written summary of her work, when offered an open invitation to write down what she had found out.

> Two even numbers added together always make an even number, because each one divides by 2 exactly and so when they are added together they will still divide by 2 exactly.

In passing, note the distinct uses of 'two' and '2', but also the passive mood and future tense as well as the disembodied frame of the whole account.

The second style, called *mixed*, refers to a transitional stage containing both words and certain special mathematical symbols (often for operations – familiar instances of common symbols include + and ×). Formulae in general frequently provide examples of a mixed recording style. Two instances from a mathematics textbook are the following.

> The volume of a pyramid is $\frac{1}{3}$ × area of base × vertical height.
> The volume of a sphere is $\frac{4}{3}$ × π × (radius)^3.
>
> (Brand *et al.*, 1969, p. 65)

Examples of a mixed recording style abound. The Cockcroft report cites the following medical example.

$$\text{Child's Dose} = \frac{\text{Age} \times \text{Adult Dose}}{\text{Age} + 12}$$

Another formula in a mixed style is:

$$\text{Water content} = \frac{\text{mls of water extract}}{\text{wt. of grease sample}} \times 100\%$$

A thirteen-year-old pupil Clare working on the following problem provided a further instance of a mixed style of recording on the way to becoming symbolic. The initial prompt to investigation was:

$$10 \times 12 = ?$$
$$11 \times 11 = ?$$

At the end of a considerable period of discussion and exploration, she was asked to say what she had shown about consecutive numbers. (I is the interviewer). She replied,

C: When you square the first consecutive number and you add the second consecutive number plus one . . . no, that's not right – when you square, you take the first consecutive number and square it . . . Oh dear. Well, if you take the first and third consecutive numbers and multiply them together, then you get the same as the middle one multiplied by itself . . . plus one.

I: Which side is plus one?

C: If you take the first and the third consecutive number – numbers and . . .

I: Do you want to write this down?

C: (Writes) c = consecutive
(Writes) The first c × third c = the second c^2 + 1

Subsequent discussion on the basis of the written record served to clarify which term was in fact 'plus one'. Here, the succinctness of the written record seemed to aid the formulation of the correct generalization.

The third style I call *symbolic*, and it refers to the prevailing style of professional mathematical writing today in which very few recognizable words appear at all, the entire text being predominantly a display of symbols from various alphabets, punctuation devices deployed with different meanings and specially invented symbols, denoting operations and relations of various kinds. No entire mathematical text is completely symbolic. For instance, even the London Mathematical Society's style sheet for potential contributors to their prestige international journals for professional mathematicians includes a request that preferably no page should be completely devoid of words.

The following generic activity (referred to as 'Borders') can be useful in producing expressions of generality which reflect these recording styles.

Given a range or sequence of related shapes made of unit squares (e.g. rectangles or 'L' shapes), put a border one square deep all the way round. How many border squares will be needed? (See Mason *et al.*, 1985.)

A thirteen-year-old pupil working on 'Borders' (with rectangles) offered the spoken description, 'Two lots of the length and two lots of the breadth and add on four', as his way to calculate how many border squares were needed for any rectangle. When asked by his teacher whether he could write that down, he spontaneously inquired, 'Can I write L for length?'. On receiving an affirmative reply, he wrote:

$2 \times L$ and $2 \times B$ add 4

Unsolicited, he then changed it to:

$2 \times L + 2 \times B + 4$

When asked how someone would know which calculation to do first, he altered his representation to:

$$2 \times L| + |2 \times B + 4$$

At the end of this exchange, he had produced a completely symbolic record of his formula.

I asked Christopher (aged twelve) to record the method he had discovered for the number of border squares. He first exemplified his method on a particular rectangle (5 × 4). He wrote:

Method

$5 \times 4 = 20$	$42 - 20 = 22$
$7 \times 6 = 42$	

I then asked him if he could write it out for any rectangle, and he then expressed his method as follows:

Length × width of first box = ?
Length + 2 width + 2 = Length × width = ?

He then paused and added:

$$
\begin{array}{r}
2\text{nd} \ ? \\
- \ 1\text{st} \ \ ? \\
\hline
= \\
\hline
\end{array}
$$

His work involves expressions of generality. Yet it is mixed in style and employs a question mark as an algebraic symbol, one which links clearly by meaning with the unknown. Christopher also developed a way of getting round the problem of his two different uses for the same ? symbol.

In this section, I have tried to exemplify different positions on a continuum of recording styles, moving from completely prose accounts at one extreme, to completely symbolic ones at the other. Before going on to expand on the move to increased symbolism signalled by the remark 'Can I write *L* for length?', I want to look briefly at the topic of precision and accuracy of records.

3 Purpose, adequacy and precision of written records

The justification offered by many teachers for a move towards the symbolic is on grounds of brevity, with mathematical notation seen as a form of shorthand. Below is a short extract from a situation where a teacher is working on functions and has encouraged her class to find a number of different ways to get from three to seven and then to express each way as a general rule. The reasons for wanting a more compact form of expression are justified only in terms of brevity.

P: Is it prime numbers? Three is prime, miss one out, seven.
P: Add on four each time?
P: Ten take away three?

The teacher then requested:

T: On the piece of paper, I'd like you to try and write down these rules first in words, then in mathematical shorthand. For example, if I wanted to write x squared and don't want to be bothered writing down s-q-u-a-r-e-d, what can I do so it's quicker?
P: x with a little two, Miss.
T: And where's the two going?
P: Above the x.
T: Right, so that's what is wanted. First write it in letters and words – in English – and then try to put it into that sort of shorthand.

Here, now, is an example of pupils on the way to a more succinct written record of observations that they have already explored verbally in detail. James and Mason (1982) provide an instance of recording by two twelve-year-old boys, who have been working on the following problem involving Colour Factor rods. As with Cuisenaire rods, Colour Factor impose an ordering and structure on the colours involved by means of the underlying length relationships among the rods themselves. The colours, used as a naming system, provide a means of recording facts about the rods' interrelations.

(Colour words never appear with numbers in the same expression, except where the numbers are used as operators (e.g.

2 Reds). This is despite particular rods occasionally being referred to by their relative number equivalents, commonly using white (the shortest rod) as the unit. This latter practice destroys one of the great potentialities of the system, namely that any given rod can be taken as the unit and then the mutual length relations determine the 'numerical' values of the others. There is a colour system for naming particular rods, so, as the colours have been used consistently, I can say that if a brown is taken as the (temporary) unit then a red represents a half and a white a quarter. The relative nature of numbering, whether by whole numbers or fractions, is lost if the rods are assigned permanent numerical names.)

> Task A Choose a rod and make up the equivalent length using repeating patterns (an example was given).
>
> Task B Talk about the rods in each row in as many ways as possible (several variations were discussed).
>
> Task C Write down the various ways of talking about the rods . . .

Some of the results they obtained for the purple rod were described orally as 'Pink-and-white . . . (pause) . . . four times', 'four pink and four white' and 'four . . . (pause) . . . white-and-pink'.

Various results were recorded as follows.

$$
\begin{aligned}
4 \text{ Pink and White} &= \text{Purple Rod} \\
4 \text{ Pink and 4 White} &= \quad '' \qquad '' \\
\text{Pink and White} \times 4 &= \quad '' \qquad '' \\
\text{Pink} + \text{White} \times 4 &= \quad '' \qquad ''
\end{aligned}
$$

Later activity and questioning raised the problem of ambiguity and the need for agreement in interpretation. It was clear from the account provided that these particular records were accurate and reinterpretable, conveying information to the participants in the situation. However, James and Mason report that when 'the teacher asked them questions such as, "If I read Pink + White × 4, what rods would I put out?", all agreed on one pink followed by four whites' (p. 251).

Punctuation devices, such as *brackets*, *underlining* or *hyphens*, are ways of making visible certain aspects of spoken language (e.g. stress, intonation, pauses). Of these, the most important for mathematics is brackets, which represent in writing information about scope which is conveyed orally by other means. The above written records reflect what had been spoken, but some of them fail to incorporate the information conveyed by phrasing and pauses. This absence is but one of the problems resulting from merely inviting pupils to write down what they have just said. Ensuring that a written transcription conveys the same message as a spoken version requires a considerable skill and linguistic sophistication.

Booth (1984) has found that, in many instances, secondary pupils working on algebra are not unaware of brackets. They merely do not see the point of using them. She succeeded in eliciting their use in written records as a consequence of operating with a cardboard 'algebra machine', which 'required' their use. For their own part, however, there seemed little perceived need by the pupils for brackets in order to render unambiguous certain mathematical expressions (even arithmetic ones such as $4 + 2 \times 3$). Apparently it is not just a matter of forgetting to put them in, and may well have to do with a lack of appreciation of the requirements of disembodied writing. Possible further explanations for this lack of awareness on their part include the following.

> There is a dominance of left-to-right processing (as with reading) which is emulated by certain types of calculators, although this can be overriden by the context.
> There is seldom ambiguity for the pupil in a particular problem because knowledge of the situation can dictate how the expressions are to be interpreted.
> There is a prevalent belief among pupils that the order of carrying out operations does not affect the result of a calculation.

The computer provides another context where pupils are willing to accept the need for brackets. Here the constraint is imposed by the machine where questions of interpretability are far less flexible. If this is what *the machine* requires, then pupils seem willing to go along with it in order to gain access to what the computer is able to offer, and brackets are a rather minor syntactic peculiarity in comparison with some which are required.

But the need is on the part of the machine, not themselves.

There are many conventions against which to judge the adequacy or otherwise of a written account. Adequacy depends crucially on the task, and the ease and clarity of communication obtained by employing a mixed or even verbal form may well outweigh the benefits of a more symbolic record. It is worth encouraging movement toward a more succinct notation, provided the context justifies it, in order to increase the pupil's symbolic repertoire, and to provide access to the ease of transformation that succinct records afford. One should not be a substitute for the other. Ideally, it should truly reflect a continuum, to all parts of which pupils should have access. They can then choose an appropriate style, appropriate both to themselves and to the task in hand.

There are major differences between conciseness and preciseness. Any of the above styles of recording (verbal, mixed and symbolic) are potentially legitimate, in that they can accurately represent what it is desired to record. Adequacy depends crucially on purpose. If my purpose is to explain to someone else what I have discovered, perhaps to someone who was not there, I might well choose to record my results in such a way as to retain the flavour of the construction. Thus, in the case of the problem of 'Borders', the pattern might be exemplified on a particular example, but structured in such a way as to bring out the general. Thus, one pupil wrote

$$\{7 + 7\} + \{4 + 4\} + 4$$

for the number of border squares round a 7×4 rectangle.

One criterion for the adequacy of a recording might be: can I go back from the expression to the perception (the 'seeing') which gave rise to it? This is quite a different criterion of adequacy from the more familiar one of the 'tightest' expression, that is, one involving the smallest number of symbols. The term used for this latter criterion is *simple*, although the notion of 'simple' form is more complex than this. Such a formulation is often sub-minimal in that it is hard to comprehend and may involve returning to the beginning, to statements such as 'let x be . . .', for referents for the various symbols.

This latter aesthetic for symbolic expressions has certain components:

  (i)   brevity over clarity;
  (ii) formal over informal;
 (iii) symbolic over verbal;
 (iv) disembodied style.

Algebraic formulations tend to be very succinct in form and therefore easier to manipulate, as they are particularly low in 'noise'. However, this compactness of form can prove a hindrance to comprehension, and such a style is not always the best method of recording. The choice of style of recording should depend crucially on the use which will be made of the records.

There is also a tradeoff between precision and expression in mathematics, exactly as there is between formal aspects of language structure (such as grammar and spelling) and creativity in expression with natural language. Even if an expression is not precise, it is worth something and should be worked on to improve it, rather than being abandoned in favour of an alternative one.

One difficulty commonly found with mathematics as a whole is a failure to comprehend what might be called 'the symbolic act'. 'Can I write L for length?' is one such instance. It is more than mere naming, although even at this level, the importance of an appropriate name is poorly understood. In her investigation of $10 \times 12 = ?$, $11 \times 11 = ?$, Clare's original notation for three consecutive numbers was $1x$, $2x$ and $3x$. She later switched to three *consecutive* letters (x, y and z) and then to x, $x + 1$ and $x + 2$. One of the conclusions she came to about what she had shown is that x, $x + 1$, $x + 2$ *is* an appropriate notation.

I: Can you say in words what it is you've proved?
C: Mm. That x, $x + 1$, $x + 2$ works and that it is the same as using x, y, and z.

We name things for reference, and hopefully for ease of reference, to draw attention to the thing named. But naming also classifies and hence causes us to look at the named thing in particular ways, the chosen symbol stressing some and ignoring other attributes of the named object. Naming something gives us power over it, particularly in algebra, as we can transform and combine expressions involving the unknown – to find out more about it.

It is incorrect to allocate precision to the language rather than the user, as is so often done in statements about mathematics. Certainly all languages have particularly well-developed registers or semantic fields in some areas, according to the demands placed upon the language by its users. Any language can be used accurately and carefully or sloppily and imprecisely. Distinctions are made according to context and intention, and precision and accuracy are functions of a user of a particular language. Precision in certain aspects is the mathematician's preference, not a characteristic of mathematical language.

Nor is it essential for all distinctions to be overt. As I mentioned in Chapter 4, the fact that Russian does not distinguish between the definite and indefinite article does not imply that the important mathematical idea of distinguishing 'there exists a unique . . .' from 'there exists . . .' is inaccessible to Russian speakers. Similarly, the fact that verbs in Mandarin are not marked for tense does not mean that Mandarin speakers cannot distinguish past from present. If a distinction needs to be made in a language, a locution, or perhaps a circumlocution, can be found.

Mathematical symbolism achieves its compact form, in part, from systematically blurring distinctions between things. Distinctions may be able to be made more succinctly by employing mathematical terminology, but it too can be employed in a vague or ambiguous manner. I can be an imprecise speaker of algebra, for instance, by failing to specify the scope of a variable. There is an important distinction between vagueness and ambiguity. Algebra thrives by being vague about the scope of many expressions. Are 2.00, 2 and $\frac{1}{2}$ equal? It depends on how we choose to see them – on the context of their use (see Adda, 1982).

There is one context of overwhelming importance in later mathematical work where the 'conciseness of expression' aesthetic is defensible, and that is where the expressions which are being formulated are to be manipulated. The more concise the formulation, the easier it is to manipulate. However, this criterion often does not apply. The superiority of concise symbolic statements over natural language formulations is unequivocal only under such circumstances, and its defects are often crippling on other occasions. It is not always advantageous to symbolize: for instance, when it is premature for the pupil or when the problem does not warrant it. The much-vaunted power of mathematical symbolization is much less apparent than is claimed. Premature symbolization is a common feature of mathematics in schools, and has as much to do with questions of status as with those of need or advantage.

4 Succinct records and the nature of algebra

The process of abbreviating words by radical truncation to single letters was briefly mentioned in the context of a move towards more succinct notation. This process takes words from a particular language and so the residual letter (presuming this process to have occurred in an alphabetic language) is guaranteed to be so interpretable only within such a language. Thus, for example, using the letters *l* for length or *h* for height provides a sort of algebraic representation with a built-in mnemonic device for interpreting the variable. It projects the mathematical notation *into* the language of the reader, and affords the pretence that the mathematics *is* written in that language.

It is an extremely common feature of algebraic notation, for good reasons of memory, that the choice of letter employed as a symbol serves as a link, not to the object itself, but more obliquely to the *word* for the object. However, the desire for common international conventions and symbols can fall foul of this pseudo-semantic link. While it is very widely accepted that letters from the Roman alphabet are to be primarily used in mathematics, whether or not they are used in the writing system of the native language (which may not even be alphabetic), if this link to words in a particular language is to be retained, the *choice* of letters must vary.

Fortunately, some textbook writers in developing countries are aware of this difficulty. A clear example of this is provided by Mmari (1975). In an attempt to develop a Kiswahili register for school mathematics, a set of terms for parts of circles (such as *centre, circumference, radius*) had to be developed. The words for *circumference* and *diameter* are *kivimbe* and *kipenyo* respectively, and there is no separate term for *radius*, the expression *nusu kipenyo* – half diameter – being employed. Consequently, the formula for the circumference of a circle was originally written $K = \pi k$, where capital and small were the only distinctions between the two k's standing for different concepts. The direct link of letter to single word no longer directly holds, and it is harder to read the formulation as an abbreviated Kiswahili sentence.

Experience showed that this was both potentially confusing and open to many errors. Two subsequent revisions have taken place. In the first, they adopted the use of the first letter of another term *mzingo* (a general term meaning *perimeter*) for the circumference and so the formula then read $M = \pi k$. In the second revision of the nation's texts, the following features were included. 'Distinct letters are used to avoid ambiguity and confusion. The formulae are standardised to conform to international practice' (p. 42).

Abbreviations of words have been regularly employed as mathematical symbols. In Attic Greek times (c. 200 BC) the numeral system used the first letter of the Greek word as the symbol for the number. Thus Π (representing the first letter of the Greek word *pente*) stood for five, Δ (representing the first letter of the Greek word *deka*) for ten, and so on. Three further historical examples of this process were the use of *ae* (Latin *aequalis*) for 'equals', *R* (Latin *radix*) for 'square root' and *co* (Italian *cosa* meaning 'thing') for 'unknown'. Such abbreviations are language-specific in that, although Latin served as the *lingua franca* of scientific discourse for many centuries, an increasing number of works were written in or translated into the vernacular. Thus 'unknown' would be *res* in Latin, *chose* in French and *Coss* in German (Boyer, 1968, p. 305).

A modern example of this process of abbreviation is the widespread use of Z for the set of integers. This arose originally as a contraction of the German word *Zahlen* (meaning 'numbers'), but as a result of the international nature of mathematical symbolism, about which I shall say more later, its use transcends German-speaking countries. The ability to reconstruct the term of which it is an abbreviation, however, depends on the reader having knowledge of both German and the symbol's origin. Otherwise it will have to remain an arbitrary symbol.

The provision of single-letter names for variables (e.g. *b* for base, *h* for height) retains a link between the new symbol and the familiar one (the word) and hence invites the algebraic use of letters to be seen generally as referring to the words. This can lead to a situation where whenever a letter is encountered, the immediate reaction when searching for a referent is to ask 'what is it short for?'. 5y is really only a shorthand for five y(achts) or five y(oghurts) (Booth, 1984). The letters are seen as units, things to be counted, nouns of which the question 'how many?' can be asked.

It is worth bearing in mind that one of the standard algebraic confusions arises in part because of the mathematician's habit of turning adjectives into nouns. Three onions becomes abstracted to the number *three* which itself has properties, e.g. odd and prime. Only when we talk about 2 + 3, rather than two apples and three apples, does arithmetic proper begin. But when we introduce measurements we write

3m for three metres,
3p for three pence,

and three has turned back into an adjective. It is not at all surprising that pupils see 5y as being another example of this type. Usage in other school subjects reinforces this means of interpreting letters. In physics, 5N stands for five newtons and 2g presents two interpretations, 2 grams or 2 'gees' (acceleration). $F = ma$ can be viewed as merely a shorthand way of writing 'force equals mass times acceleration'.

There is a type of problem which brings out questions of word order and the difficulties of interpretation of single-letter variable names.

In a certain school, there are 30 pupils for every teacher. If T represents the number of teachers and P the number of pupils, write down an equation relating P and T.

It is very easy (but incorrect, given the specification of these variables in the problem) to write $30P = T$. This equation reflects the order of occurrence of the words in English and 30 certainly seems to refer to P. An equivalent situation comes from writing $3f = 1y$ as a symbolic rendering of the proposition 'there are three feet in one yard'.

One of the major difficulties for many pupils is trying to come to terms with the way letters are being used. A common perception seems to be that of letters being objects, or a shorthand for objects, in their own right. 'Fruit-salad' algebra is a common term for the practice of interpreting, for example, $5a + 2b$ as five apples and two bananas. This is sometimes offered as an attempt to forestall the collection of symbols (5 add 2 is 7 and an a and a b, so the answer is $7ab$) and permit $5a + 2b$ to be seen as an entity in its own right. Unfortunately, it leads to confusion between a being apples (under the same truncation process as length becoming l) and a being 'the number of apples'.

The distinction is between the objects themselves and the number of them. The algebraic expression $5a$ is not an analogue of five apples, nor is five apples a possible interpretation of $5a$. Apples are units which can be counted just like grams or centimetres whereas algebra generalizes the numbers (the coefficients) and the letters themselves are standing for *numbers*. Unfortunately, using the union of sets as a model for addition can carry over into the symbolic world, where the members of the sets are seen as letters, and rewriting $x + x + x$ as $3x$ seems the same as collecting counters together. There are even teacher-offered instructions such as 'collect like terms'.

At a more subtle level, there is the belief in mathematics that pupils should be aware of the arbitrariness of the choice of letter name of the variable. Therefore this mnemonic naming system is not rigidly adhered to. On the contrary, it is conventional for many variables to be called x or y independently of the referents. It is perhaps an irony that although x is the letter which in English permits the fewest expansions into object names, it is the use of this letter which is taken as the hallmark of algebra. Or was this why it was chosen?

What is algebra?

In general, there is one major drawback to this system in any language, namely the danger of the symbol being interpreted as the object itself. To illustrate what this might mean, below is part of an interview between myself (D) and two fifteen-year-olds Sharon (S) and Christopher (C), concerning the question 'What *is* algebra?'.

C: Well, it's substituting letters for numbers in order to work out an equation.
S: (laughs) That sounds about right. Using letters instead of numbers.

D: Is it anything else?

S: It's a pain in the neck.

D: Uh-huh. Why is it a pain in the neck?

S: It confuses me.

D: Do you know why?

S: I suppose, because you're used to using numbers. Then it should be easier really, but . . .

D: Why should it be easier?

C: Because you're to think of letters as numbers and treat them in the same way.

S: Yes, I don't understand though. It's where you can add them sometimes but you can't . . . like our teacher says about adding apples and pears, that you can't add apples and pears and I don't know what he is talking about . . . because if it is a number you can see whether you can add it or not, can't you.

D: Why do you think he goes on about apples and pears?

S: Because it is something we know I suppose. To get you to try to do it, but I still don't understand. It's because you can add them together when you're timesing them, but you can't add them together when you're plussing them or something. It doesn't make sense to me.

D: I'm interested in this adding and multiplying apples and pears.

S: I can't . . . I don't know what it is, but you can't add a to b. He said you can't add apples and pears but you can multiply them and I don't get that. I mean, you can't multiply apples and pears, can you?

C: Because you can imagine an apple being an x and a pear being a y and if you add apples and pears you get sort of a pearapple or something.

S: Yes, you can get fruit though.

It is not clear that even following the lead of computing, where variable names are frequently longer, e.g. SIDE, will solve this difficulty of interpretation. In the case of SIDE, it is still unclear whether or not it is the length of the side, the number of sides or the side itself which is being referred to. Thus the same symbol may be used, in different contexts, for meanings which are in conflict.

Clare had just spent 20 minutes working on a problem

involving a sequence of shapes and tried to find a way of expressing the pattern that gave the number of squares which made up any particular shape in the sequence.

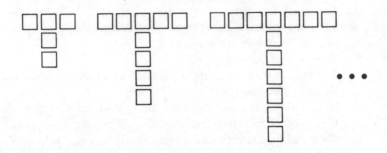

Diagram 8

She finally wrote her conclusions as follows.

The number of squares will always be an odd number.

x = number in sequence
$x \times 2 + 1$ = no. of squares on horizontal row = y
$y \times 2 - 1$ = total no. of squares needed = z

This record as it stands reflects the sequence of events she undertook in coming to the final formulation. Once tidied up algebraically, perhaps in the concise form $z = 4x + 1$, its genesis is no longer recoverable from the record. It may no longer be possible to go from the record back to the *seeing* of which it is ultimately a record, in order to see *why* the above answer is correct.

I then posed Clare the question 'What is algebra?'

C: Hm. It's substituting numbers that you don't know for a letter. You can use it to make a rule for, um, for any number. Or you can use it in different ways. It's just using letters instead of numbers.

D: What is the advantage of using letters?

C: Well, if you don't know what the number is, or if there is a whole set of numbers, it would be quicker if you used letters to make a formula, and then just use it on the whole set of numbers.

D: Is what you have just done algebra?

C: Well, . . . I don't know, because, um, I only used it instead of – I suppose it is algebra, but I only used it instead of words because it's quicker. Writing words or a letter doesn't really make a lot of difference.

D: Do you find it easier writing words or letters?

C: Letters. It makes it look easier.

D: Is it easier?

C: It makes it quicker to write. But I wasn't using it, sort of thinking of it instead of the number.

D: I see. So this isn't really algebra.

C: It is algebra, but I just wasn't thinking of it that way.

D: How were you thinking about it?

C: I was thinking if you didn't know the number then I was just writing what sort of number it was and then I was writing a letter instead of that.

D: What else could you have done?

C: I could have just put a letter in the first place, if I had known what sort of a number it was going to be.

D: What do you mean by 'what sort of a number'?

C: Whether it was like, the number in the sequence or the number of squares or the number in the horizontal row.

D: These were different things you were working with?

C: Yes.

The above transcript suggests a number of questions about symbolic representation. The main one I wish to mention here is that the choice of a notation that is both suggestive yet clear is of utmost importance, and in algebra, there are a number of subleties in this area which can lead to definite difficulties. In particular, an approach which involves naming the key variables at the outset, for example by letters, requires you to know what those variables are.

A concern with the *form* of algebra (use of alphabetic symbols, particularly x and y) dominated many of the recollections of adults whom I have interviewed. The following is a response to a question about the nature of algebra. There are considerable shifts in precisely what the letters are standing for as the conversation continues.

For example, call o oranges and p pears. Well o is a bad letter because you might confuse it with nought. Let's use r. You could have raspberries if it is r. It doesn't matter what letter you have. If you used a or b though, you might not known whether b was pears or oranges. Suppose you had a pound of oranges and three pounds of pears. So you might have $1r + 3p = 40$ pence, the total cost of the bill. $1r$ is the value of 1lb of oranges, r and p are different values. So it is really price of $1r$ plus price of $3p = 40$. Suppose $r = 4$, r is a price.

Even the *Concise Oxford English Dictionary* (Sykes, 1982) donates equal space to form and purpose in its definition of algebra as 'a branch of mathematics dealing with properties of numbers and quantities by means of letters and other general symbols'.

Algebra, in the sense of the formulation and manipulation of expressions of generality, is perfectly possible without letters, for example by using words or non-alphabetic symbols, e.g. a cloud or a box. It seems totally unnecessary to tie algebra to alphabetic notation (as testified to by the increasingly prevalent usage of box, square and triangle as geometric icons for variables). But in partial answer to the question 'What is algebra?', one offered at a far more profound level, let me end by quoting Hewitt (1985, p. 15).

Algebra is not what we write on paper, but is something that goes on inside us. So, as a teacher, I must realise that notation is only a way of representing algebra, not algebra itself.

Summary

Written mathematics is clearly not just spoken mathematics written down in words. In this chapter, various alternative styles of recording mathematical ideas and generalizations have been exemplified. The historical fact of a move toward the highly symbolic recording style of contemporary mathematics can be juxtaposed with the movement toward succinct notation observable in pupils as they progress through school. There is a widespread feeling that somehow only the fully symbolic representations are mathematical, and there is a strong tendency for teachers to move quickly to, for example, single letter variables.

All the examples I have provided of written and printed mathematics attest to the widespread use of letters of the alphabet used in unusual ways and as non-alphabetic symbols. There are various forms of recording available, all of which are potentially legitimate if they are used accurately.. There are many advantages of symbols over words as a medium for recording mathematical ideas, but the case is not completely clear-cut. In particular, pupils frequently fail to have a clear idea of *why* they are recording and, without any feeling for the purpose, it is difficult to discover what, for example, is ambiguous or insufficient in some way.

In conclusion, considerable attention needs to be paid to questions of how children record mathematics spontaneously, and what they find worthwhile to record in a particular context where both the purpose and the need to record are clearly imposed by the constraints of the situation. In particular, what are the purposes to which disembodied language in the form of written records is put, and how might these purposes be conveyed to pupils? What do pupils find useful to record? Is the audience clear and known? Is the purpose known? What conventions are operating which govern the form in which records should be written? These and many other questions seem to me to be central to an understanding of the place of writing in mathematics.

6

Some features of the mathematical writing system

We are tied down to a language which makes up in obscurity what it lacks in style.

Tom Stoppard, *Rosencrantz and Guildenstern are Dead*

I pointed out in Chapter 1 that the symbolic aspect of mathematics was one of the subject's most apparent and distinctive features. Skemp (1979, 1986) has written extensively on the variety of functions performed by symbols in our highly symbol-dense environment, where his use of the term *symbol* includes both spoken sounds and written words as well as many others. Words are symbols, but ones which fall into a special category in that they are so familiar and commonplace that they come strongly to stand for what they are symbolizing. For example, if the word *justice* is read, certain images and associations immediately come to mind.

Purpose	*Comment*
Communicating	Symbols allow access to the thoughts of others
Recording and retrieving knowledge	Special cases of communicating
Helping to show structure	Relations between the symbols reflect relations between the ideas
Allowing routine manipulation to be made automatic	Freeing conscious attention for other things
Making reflection possible	Putting thoughts 'out there', with some stability, compactness and permanence, as objects which may be studied

Above (adapted from Skemp, 1979) is a list of the uses to which symbols can be put which are of particular relevance to mathematics teaching and learning. Certain aspects of these various purposes with regard to written mathematics have been discussed in the previous chapter. The concern of this chapter is entirely with the third and fourth items in this list, in particular with the symbol systems used in mathematical writing and how they are structured.

It has often been remarked that it is important to be able to distinguish at will between the symbol and the concept, between the signifier (the symbol) and the signified (the referent). The symbol is not the referent, although it is normally functioning best when it can substitute strongly for it. In mathematics, one major use of symbolization is precisely to allow manipulation to move faster and more seamlessly by blurring the distinction between symbol and object. However, both the signifier and the signified have their separate attributes.

To be truly symbolic, however, there should not be an iconic relationship between symbol and referent. In an article entitled *When is a symbol symbolic?*, Mason (1980) discusses his view of symbols as windows, rather than mere marks on paper. One sees *through* symbols if they are functioning correctly, and with this perspective it makes sense to ask which seeings they afford and which they block. It may well be of benefit to have a number of different notations for the same concept, provided each highlights some and ignores other aspects of the related concept. For instance, some notations emphasize by making certain distinctions visible, while others do it by means of connecting via visual similarity of form. For example, each of $Df(x)$, $f'(x)$ and $\frac{dy}{dx}$ is currently employed to stand for the same notion, namely the derivative of the function $y = f(x)$.

As a result of making language visible, a finer control and discrimination mechanism is available, namely the visual/manual rather than the aural/vocal. 'One cannot help wondering whether man's optical faculty does not endow him with greater powers of discrimination than does the acoustic' (Pulgram, 1976, p. 18). In teaching mathematics to blind students, complex spoken expressions are required in order to convey all the structural information which is coded conventionally by means of spatial position alone. The expressions a^{b+c}, $ab + c$ and $a^b + c$ employ identical symbols appearing in the same left-to-right order, yet each means something quite different from the others due to the positional variations exploited by the mathematical writing system. The basis of *systematic representation* is to relate the symbol and meaning in a principled way.

Many languages have a writing system and a syntax of written forms; so does mathematics. Writing systems exist somewhat independently of languages, and certainly do not by themselves constitute languages. (So the mere fact that mathematics has a complex and sophisticated writing system does not entail it being a language.) For instance, English and French employ the same writing system, i.e. the same range of symbols are used in the written form of the language. Diacritic marks (such as accents) are used to adjust the writing system to the particular language.

My intention in this chapter is to examine both the range of symbols which constitute the mathematics writing system and the conventions that are used to discriminate meanings. The first section will examine the basic symbols themselves, while the second will indicate the limited ways available for combining such symbols into compound symbol clusters employed to express mathematical relationships between the component ideas. The final section examines the confusions which can (and need to) occur between symbol and referent in mathematics.

1 What symbols are conventionally employed in mathematics?

Is there a basic set of symbols from which all mathematical notation is built up? 'The mathematician manages with ten ciphers plus about the same number of functional symbols and a few alphabetic letters on occasion' (Brice, 1976, p. 43). It is clear from looking at a page of mathematical symbolism that, while there is some truth to Brice's claim, his is a considerable underestimate. It also fails to illuminate the structural relations which appear to govern the particular choice of alphabet, specific letters or even whether a capital or a small letter is employed. It is as well to point out at this stage that, despite a widespread use of letters from a range of alphabets, mathematical writing as a whole is not alphabetic in character. I mean this in the sense that

English has an alphabetic writing system whereas Mandarin Chinese does not.

I wish to separate the symbols used in mathematics into four main classes: *logograms* (specially invented signs for whole concepts), *pictograms* (stylized icons in which the symbol is closely related to the meaning), *punctuation symbols* and *alphabetic symbols*. I now look at each class in turn and make some remarks about the range and type of use to which it is put.

Logograms

There is a collection of special symbols, invented shapes which are not used outside a mathematical context. In this sense, they parallel (but do not always represent) the *technical* terms from the mathematics register. These fit the description of *logograms*, that is, special symbols which stand for whole words. Everyday language examples of logograms include '$' for *dollars* and '&' for *and*. The range of special mathematical symbols is a little wider than Brice's estimate of ten cited above. The most familiar instances are the numerals that Brice termed *ciphers*, namely 0, 1, . . ., 9. These do not have a capital or small variation, although there are some slight variations in their formation, e.g. the 'continental' 7 and corresponding 1, or their computerized presentations. Different symbols for these digits are employed in certain parts of the world, e.g. Arabic or Thai numerals. (It is a minor historical irony that the so-called Arabic numerals, 0, 1, . . ., 9, are not widely employed in most Arabic-speaking countries.)

Other examples of mathematical logograms include $+, -, \times, \div,$ %, $\sqrt{}, -, |, =>, <=>, \therefore, \in, \cup, \cap, \subset, \int, \cong, \circ, \vee$ and $\wedge$. All of these symbols are referred to as *signs*, such as the square root sign and the integral sign. The latter provides an interesting example as it started as a capital letter, a one-letter contraction of the Latin word *summa*. Over the centuries, its shape and size have altered and it has become an identifiable logogram on its own. The pounds sign is a comparable embellishment of a capital Roman L from a similar abbreviation of the Latin word *librum*. Non-recognition of Σ or Π as Greek capital letters means they are likely to be seen as logograms. If so seen, the structuring implicit in the choice of first letter (s for sum, p for product) will be lost.

A converse situation can be seen with the set-theoretic symbols for Union and i∩tersection. A false English folk etymology has been attempted by means of the above to act as a mnemonic device to link the symbols to the English word for their meaning. The widespread use of the principle of first-letter contraction renders this being the actual derivation of the symbol for intersection unlikely. Other examples include already cited symbols in an altered, stylised presentation and orientation. Backwards E(∃) and upside down A (∀), first letter contractions of *exists* and *all*, but in an altered orientation, are used as logograms by the majority of mathematicians.

Another common example of such a symbol is the '=' sign. It had an iconic origin in the sixteenth century, as its coiner R. Recorde (1557) indicated. He wrote,

> I will sette as I doe often in woorke use, a paire of paralleles, or Gemowe [twin] lines of one lengthe, thus: =====, bicause noe 2. thynges, can be moare equalle.

This explanation of the choice of the symbol shows a marked pictographic element in Recorde's mind, and also serves to indicate that my proposed categories are not hard and fast. As a further instance of this, while the logograms > and < (for inequalities) are suggestive of their meaning, they are not strictly pictographic either.

Pictograms

There are a few geometric icons used in mathematics, *pictograms* to be precise, where the symbol is a stylized but clearly interpretable image of the object itself. For instance ∠ for angle, □ for square, ⊙ or ○ for a circle and △ for triangle. They are symbols in that they are stylized representations; for example, the triangle is always equilateral and in the standard orientation with respect to the page. Another composite adapted symbol with a pictographic element is $\oint$ for contour integral.

Not all single-letter naming is alliterative. One exception is provided by the geometric terminology of Z angles and F angles. Other common names for these angles are *alternate* and *corresponding* respectively. Why Z for alternate and F for corresponding? The origin is in fact pictographic, whereby the closest letter has been chosen which reflects and emphasizes the configurations of parallel lines and a transversal giving rise to the angles themselves.

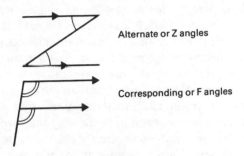

Alternate or Z angles

Corresponding or F angles

Diagram 9

There is a more complex problem with diagrams, especially those of geometric figures. In some circumstances, a drawn rectangle, for instance, appears to be the actual object of study, and measuring it with a ruler and protractor is an appropriate activity. On other occasions, the same configuration of lines can be used as a pictographic symbol, where the actual dimensions, orientation, etc. are irrelevant or conventional. It is possibly being employed representationally to stand for a particular rectangle, or perhaps generically for them all.

Punctuation symbols

Many symbols ordinarily used for punctuation in standard English orthography are also widely employed in mathematics, though not commonly for punctuation. For instance, the colon in a:b is used to denote the *ratio* of a to b (the use dates back at least to 1651). It also serves to punctuate the definition of a function in mathematics, e.g. $f:A \to B$, as well as a separator in the description of a set, e.g. $\{x : x>2\}$. (It is also used to denote a repeat in musical notation.) As a second instance, ';' is used to denote the operation of partial differentiation in tensors. Further instances include: ',' and '.' (decimal point in France and Britain respectively), '!' (factorial), (, {, [(and their pairings), '∗' and '/'. (Notice how use of quotation marks allows the symbol itself, rather than its conventional interpretation, to be the focus of attention as an object in its own right, and is particularly necessary in a context where the same symbols are being conventionally employed.)

Each of these punctuation marks has an English word to name it, e.g. *comma* or *colon*, which allows it to be read at the level of punctuation device, rather than in its mathematical use. (I shall discuss some aspects of reading symbolic mathematics in Chapter 8.) These symbols, like letters, are commonly available (e.g. on typewriters and print fonts) for employment in the service of mathematics (or anything else). Diacritic marks such as ^ and ' are also employed in a similar fashion to the way they are in standard orthography; not as symbols in their own right, but to modify already existing ones. One instance can be seen with the notation $f(x)$ for a function and $f'(x)$ for its associated derived function, that is, the one obtained from $f(x)$ by the process of differentiation. Similarly, there is a general negation construction which is to add \ or / as a diacritic modification through any of a range of relational symbols (e.g. =, =>, >, ∈, ⊂), but not through any object or operation symbols.

However, at least one of the available punctuation symbols is not conventionally employed as a specialized mathematical symbol in mathematical printing, namely the question mark. It is usually employed informally in mathematics as a suggestive symbol for the 'as-yet unknown', or as a diacritic modifying an equals sign to indicate a conjecture. It carries over from ordinary language use and connotation a sense of question, a tension to be resolved, perhaps because it is so strongly linked with its meaning.

In the last chapter, I provided the example of a twelve-year-old pupil who, when designing his own notation for an algebraic representation of a general situation, chose to employ a question mark as a variable name. A first-year university mathematics student working on a problem at the forefront of his strategic, if not his technical, knowledge wrote:

$$x := \frac{\cos(\theta)}{?}, \ y = \frac{\sin(\theta)}{??}$$

The denominators in the two cases represented two completely independent (but corresponding) lengths in two different right-angled triangles. Repetition for him produced a different symbol, reflecting a non-algebraic juxtaposition, for in conventional algebraic notation ?? would stand for $?^2$. A more conventional notation, invoking ? as a variable name, would have been to call them $?_1$ and $?_2$.

This example provides a nice illustration of two conflicting features: on the one hand, the essential arbitrariness of the symbol–referent link, and on the other, how strong are the conventions guiding such choices. One of these conventions is that variables or unknowns are denoted by alphabetic letters (usually Roman or Greek) rather than punctuation symbols (which are more frequently operation symbols), so much so that the above uses of the question mark seem almost like a category mistake. It is anomalous for a mathematically-educated person, i.e. one who has been exposed to and has absorbed the guiding conventions (whether or not articulated overtly), to read

If ! = 4 and if , = 6 then find ? such that $,^2 + ? = !$

in a way that is not the case for

If $a = 4$ and if $b = 6$ then find c such that $b^2 + c = a$.

The importance of such symbol fixation is not to be underestimated, particularly as in mathematics a widespread shifting in the meaning of certain symbols (such as the '×' sign to be discussed in Chapter 8) is a regular occurrence. (In the same way, the numerals are so closely assigned to their meanings that it would do severe conceptual violence to use the number symbol '6', for instance, as a variable name.)

Alphabetic symbols

To end this collection of the available symbols employed in mathematics, there are the various alphabetic symbols. The Roman alphabet, $a, b, \ldots, z, A, B, \ldots, Z$, and the Greek alphabet $\alpha, \beta, \ldots, \omega$ (and capitals), are the most consistently and completely employed. There are occasional borrowings of extra letters from other alphabets, e.g. aleph, and occasionally bet, are the only letters of the Hebrew alphabet which have found uses within higher mathematics. Also script Roman capitals are sometimes seen, as is Gothic script. Roman letters are used for mathematics even in many countries with non-Roman alphabets. Thus the symbolic aspect of a page of Russian, Israeli or even Chinese mathematics will be virtually identical to one printed in England.

With regard to the use of alphabetic symbols, there are a number of conventions operating. Within a given set of alphabetic characters, there are conceptual distinctions marked by differences in alphabet or font, such as vectors being represeٮted by Roman letters (while scalars are Greek ones) or vectors being printed in bold or italic typeface. A comparable distinction can be seen in the consensus decision of seventeenth- and eighteenth-century mathematicians to adhere to Descartes' system whereby letters near the beginning of the alphabet are employed for parameters and letters near the end of the alphabet for variables. (The general form of the quadratic equation $ax^2 + bx + c = 0$ illustrates this convention quite well.) This was in preference to Vieta's use of vowels for variables and consonants for parameters.

The distinction of Vieta's which lay at the basis of his symbolic convention was a functional one between types of letters. Descartes' was a surface one, namely the conventional order of appearance of letters in the alphabet. My suspicion regarding the preference is that order is a far more simple and salient characteristic in the mathematical context than is the vowel/consonant distinction.

A further convention is one of employing consecutive letters for like objects, when no other factors are operating. Thus, fractions tend to be $\frac{a}{b}$ or $\frac{p}{q}$ rather than (say) $\frac{n}{d}$ (which could be alliterative naming for numerator/denominator). Once an alliterative choice has been made, this principle also allows a systematic continuation. For instance, f is a common choice of variable name for a function, and further functions are then likely to be called g and h.

I have been unable to find any consistent convention in formulae with regard to when letter variables are written with a capital or a small letter. A common consistency is for vertex and angle names to be upper-case, while lengths are lower-case, with the corresponding side in a triangle having the same letter as its opposite angle. A second consistency is to use small letters for elements of a set and the corresponding capital letter for the set itself. One counterexample to this latter regularity is provided by the reasonably widespread use of k to stand for a general field. (The common alternative form is F, which is both a capital letter and an alliterative contraction.) The letter k provides a double irony, in that it is a contraction of the German word *Körper* and all nouns in German are written with a capital letter. Thus both on German orthographic grounds as well as mathematical set-theoretic ones, one would expect this to be written as K.

My original conjecture was that as many formulae in textbooks are singled out and hence are on a separate line, then following the capitalizing rule for the beginning of a sentence in standard English orthography, the first variable should have a capital letter. Thus,

$$C = 2\pi r,$$

where C stands for the circumference of a circle and r the radius. However, there are numerous counterexamples to this simple, initial hypothesis. Certain variables seem consistently to rate a capital and others a small letter, while a third class oscillates.

Other relations across alphabets are sometimes employed to make category distinctions. In statistics, the letters m and s are used for the *mean* and *standard deviation*, respectively, of a sample, while the corresponding Greek letters μ and σ are employed to refer to the corresponding population parameters from which the sample came. A further instance, this time of alliterative first-letter links across alphabets, involves the Greek letter capital delta (Δ), which is used for the discriminant, namely the number $b^2 - 4ac$ from a quadratic equation $ax^2 + bx + c = 0$. With different meanings, $\frac{\Delta y}{\Delta x}, \frac{\delta y}{\delta x}$, as well as the notation $\frac{dy}{dx}$, can all be found in calculus texts. If the aural/alphabetic link between d and the two forms of delta is unfamiliar, then these symbol clusters may seem unrelated. Yet, although linked, subtle distinctions are being marked by the change of alphabet of related symbols in the same configuration, just as they were in the earlier statistical example. To someone unfamiliar with the aural links between the various alphabets, this level of systematic structuring will be inaccessible.

The alliterative first-letter link between the symbol used and the ordinary-language word for the concept (e.g. using m for the mean) is widespread in mathematics, and the educational ramifications of this practice were discussed in Chapter 5 in the context of algebra. It clearly depends on the language of the reader, and on occasion, as I have shown, there will be a conflict between the conventional letter employed in mathematics and an unwillingness (on grounds of easing the memory load) to employ that particular letter.

While I have left out some of the more sophisticated symbols of higher mathematics, I do not think that I have omitted any major class which is currently employed. Mathematical notation is an enormously conservative system in that the same symbols are used repeatedly with differing meanings in different contexts, rather than new ones being invented. Apart from the serious costs of printing mathematics, there is also a more general point about the functioning of symbols.

If they are not to obtrude, it is important that the symbols employed already exist as a conceptual object for the user, which is to say that they can be recognized, formed and distinguished without effort and preferably without any conscious attention. The problem of reading mathematical symbols, as well as the problem of inner articulation so they can be pronounced when self-muttering is going on, will be explored briefly in Chapter 8. In my attempt to characterize the important aspects of the mathematical writing system, I now move from the individual symbols themselves to a consideration of the varied ways in which they may be combined.

2 Which features are systematically exploited in mathematics?

The catalogue of symbols listed in the first section is merely an unordered set. To qualify for the term *symbol system*, there must be some structural relations which allow these symbols to be combined to create new compound symbols. In alphabetic languages, juxtaposition is the sole means by which words are formed from the letters, that is, linear concatenation of these elements. In ancient Greek manuscripts, there are no spaces employed as a device to separate words, and letters have only one form, so a text is just a continuous string of letters. There are many more and varied means for combining and distinguishing collections of symbols in mathematics. Unfortunately, many pupils see merely a linear string of symbols and are unable to gain access to the systematic aspects of the notation which are widely employed in order to make important discriminations and carry information. It is to this topic that I now turn.

Recording involves making certain aspects of language visible, and there are both advantages and disadvantages in so doing. Mathematical ideas are often conveyed using specialized, highly condensed symbol systems (which interact and, as I shall show, sometimes interfere). These systems attempt to reflect relationships among the ideas by means of relationships among the symbols. In so doing, the symbol system acts as a kind of filter, dispensing with all but the relevant elements involved, as by no means all the relationships among the ideas can be simultaneously represented. Also, however, extraneous relationships can appear which exist between the symbols customarily employed, but which do not reflect a corresponding relationship between the associated ideas.

Symbols provide an efficient means of storing and conveying information, because they allow the compression of a lot of information into a small space (e.g. in a formula). Symbolic mathematical expressions are also very often low on redundancy (unlike everyday speech) which can make for very dense reading, and in many instances they can prove sub-minimal in terms of interpretability by non-experts. Much of the increased precision arising from using symbols, as well as clarity and specificity of reference, comes from this quality of succinctness.

Information can be stored in many ways and, in order for it to be used, the conventions regarding the structuring must be known. This is true of any symbol system although, until it alters or is in some way brought to our conscious attention, we are often unaware of our tacit, interpretative knowledge. In the rest of this section I take a look at a number of principles which are employed on a systematic basis in mathematical writing, but in order to provide a contrast, I shall first explore a potential principle which is not commonly used.

Colour

Colour codes on electrical wires provide information, as do the lights on ships and planes, as well as the different colours used on roadsigns in Britain (blue for motorways, green for major roads). Much of the information coding which I have mentioned so far exemplifies tacit knowledge, that is, knowledge about a system which does not accompany the system itself in use. Fluent users can operate merely by means of the colours. The conciseness would be lost if interpretations of the particular symbols and information concerning how the system worked had to be provided on every occasion.

Colour is not a common functional attribute of printed words, although on occasions it is used as a decorative one. In his book *Reading with Words in Colour* (1969), Gattegno used colour as a means of directing attention towards salient aspects of words in learning to read. He has also designed a computer program to use colour and position in combination in order to represent the tonal complexities of spoken Mandarin. In Mandarin, any spoken character involves one of three separate tones, and often the three sounds have completely independent meanings.

The advent of multicolour microcomputer screen displays (as well as colour photocopiers and information systems such as Prestel) is altering the present situation. However, the current use of this new technology seems to be primarily decorative or redundantly contrastive, for example, distinguishing different parts of the text by both spacing and colour. The fact that some computers can distinguish between a blue 'b' and a yellow one, for instance, indicates that the potential for wider systematic use is there.

The systematic use of colour in mathematics is uncommon – I could find only a few mathematical instances of its use. The first two employ colour to discriminate between two different functions of particular entries. In accountancy, red is used for debit entries and black for credit entries, a practice giving rise to the expression 'in the red'. In the Rhind mathematical papyrus (and a couple of other Ancient Egyptian papyri) there are the so-called red auxiliary numbers (Neugebauer, 1969). These seem to be the by-product of specific intermediate calculations necessary for the solution of the main problem. Both these examples use a two-colour contrast only.

Colour is seldom used in printed mathematical notation, in part as it cannot be easily replicated either by the teacher or by pupils. By this I mean that there are very few instances where the same symbols are employed, differing visually only in colour, but meaning different things. A school mathematical example arises in Morocco (Daife, 1983), where it is common practice for both books and teachers and pupils to write negative numbers using green numerals and positive numbers using red numerals. Unsigned numbers are printed in black. Thus instead of −6, a green 6 is employed. No special symbolisms such as + or − modify the numbers themselves, so users of this system cannot be confused between these symbols employed as sign markers and their use for the mathematical operations of addition and subtraction. However, later fluent calculation relies to some extent on both the elision of some of these signs (e.g. writing +2 as 2) as well as the deliberate confusion of different uses of the same sign (e.g. reading −2 as either 'subtract two' or 'negative two').

Order

A common means of conveying information is by means of the order of written items on the page. Different orders for the same symbols in English change the meaning, whether at the level of letters e.g. 'bat' versus 'tab', or at the level of words, e.g. 'John loves Kate' versus 'Kate loves John.' In mathematics, for example, 17 and 71 comprise the same symbols, differing only in their order of presentation. They mean different things, which is to say that symbol order *by itself* is used in mathematics to reflect such differences.

Most writing systems exploit order to increase markedly the number of options available with a fixed collection of symbols. In much of western society we are so used to processing information from left to right, superimposed on a top-to-bottom orientation, that it is extremely hard to conceive that the order might be otherwise. Typewriters and computer screen displays produce text this way, and part of learning to read involves acquiring an automatic scanning technique in this manner. This procedure places a linear order on items displayed in a two-dimensional array, and reflects the linearity of time order in speech. Chinese and Japanese writing involve the use of many different characters and frequently proceed from top to bottom and then from right to left. Both Arabic and Hebrew script go from right to left and then top to bottom.

The Hindu-Arabic numerals (0–9) are international, but not yet worldwide. The numeration system is so dependent on order that it goes against the prevailing order of writing in many countries. The value of the digits 0 to 9 varies according to the relative position in the representation, but this is not necessarily the case with all numeration systems. A different system (Ionian) was that employed by the ancient Greeks, whereby the numerals were the letters of their alphabet. α to ι stood for 1 to 9, the next nine letters for 10, 20, . . ., 90, and the final nine for 100 to 900. To deal with numbers larger than 999, an extra notation was used (a comma) so that ,α meant 1000. Clearly, since each symbol carries its own absolute value, order is irrelevant.

This system also provides some basis for numerology, in that words *are* numbers at one level, because most combinations of Greek letters add to some value (where juxtaposition is interpreted as simple addition), although not all numbers form words. (The exceptions implied by the word *most* involve repetition of letters, which is numerically unnecessary in this system, and the implicit order of increasing value, which mark some strings of letters as symbolically peculiar numbers). Consequently, if certain words are very important in a culture, then their corresponding numbers are likely to be held to be too, as are numerical relations among words, e.g. ones which add to the same numerical value. It is merely a question of perception and context as to which is seen, word or number, when looking at those collections of symbols which admit both interpretations.

A comparable system, known as *gematria*, still operates in Hebrew, whereby the twenty-two letters of the Hebrew alphabet have numerical equivalents. At the time when I was writing this section (July 1984), it was year 5744 in the Jewish calendar. Just as in the UK where it is common to refer to the year by the last two digits alone, e.g. '84, it is common to refer to the Jewish year by the last three. 744, the last three digits of the then current year, is obtained as 400 + 300 + 40 + 4, and so its name comprises the four letters *taph*, *shin*, *mem* and *daled*. This is pronounced *tashmad*. Now *tashmad* happens to be a Hebrew word meaning 'destruction' and, as a result, many people are referring to the year as *tashdam*. Because order of the symbols is irrelevant in determining the overall value, *tashdam* and *tashmad* are (numerically) equivalent.

Position

Position, both absolute and relative, is also commonly employed as a means of representing different ideas with the same basic symbols. This frequently involves subtle discriminations in spacing and juxtaposition of symbols. Power notation and suffix notation rely on a combination of position and relative size (itself another principle) to distinguish 23 from 2^3, for example, or x^3 from x_3. Fractions are frequently written in the everyday world as backward powers, for example writing $^1_3$ m on motorway signs for one third of a mile. The common fraction representation $\frac{a}{b}$ is to be seen both as a single symbol and as one made up of components. More complex symbols involve a two-dimensional structure such as with the summation sign $\Sigma_{i=1}^{k}$ or the definite integral $\int_a^b$. Other instances include $\sqrt[3]{}$, $\dot{x}$ for $\frac{dx}{dt}$, and $\overset{\circ}{N}$ (the interior of a set N).

In Chapter 8 I will point out that most of the conventions of reading with regard to the order of processing of the symbols carry over to mathematics from English, although in the case of certain of these two-dimensional symbols there are on occasion exceptional, secondary rules of reading from bottom to top or conversely. Tensors $a_{j,k}^i$ reflect the order in which their component symbols are to be read in the alphabetical order of the customary subscript and superscript variables. Thus not only are the positional relations of *immediately to the left/right* employed, but also those of *immediately above/below*. Lying behind the functioning of this notation is the central idea of the main line of print, against which these various deviations from a linear flow of symbols are located.

Relative size

Although there are standardized sizes for all of the digits, variations in size are employed. Powers tend to be smaller than their bases, and if there were a capital/small distinction for digits, I suspect that this would be one place where such a difference would be employed. The advent of decimal pricing in garages has resulted in non-standard decimal representations, utilizing varying numeral sizes. In the past two years, I have seen only one garage that actually presented its prices in the standard decimal format, namely £1.625. The common practice is now to use a size and modified power format, thus £1.62$^{.5}$ where the size and position is used to diminish the salience, and hence importance, of the final digit. There is a residue here of a form-meaning link. It is true with regard to overall size, that for whole numbers, the more digits present, the larger the number. But the fact that seven is greater than four is not reflected in the relative sizes of the respective numerals. An airline advertisement recently illustrated this point nicely.

British Midland Heathrow–Edinburgh £xx return

British Airways Heathrow–Edinburgh £x!*!xo! return

It seems someone is trying to pick your pocket.

The inference that you are supposed to draw is that the cost via British Midland is less than £100 return (though not necessarily a multiple of 11), whereas the British Airways cost is astronomical (the implicit suggestion is millions of pounds, if you combine an algebraic presupposition of one digit per symbol with that of the familiar place-value principle). The expletive is also not far away.

Orientation

Orientation is part of what makes a particular symbol distinctive, forming a component of its identity. (Orientation is almost the sole distinction between, for example, d and p, and u and n: similarly, in mathematics, for + and ×, < and >, ∪, ⊂ and ∩, 6 and 9.) Confusions between letters themselves and/or numerals is often a difficulty for young children struggling to sort out samenesses among the bewildering array of marks which they see and produce daily. (See Higginson, 1980, for a very clear example of this struggle.) A further instance can be seen in the orientation of the symbols for 'contains' (both between sets ⊂, and between sets and members ∈) where orientation left-right is reversed to give the reverse relationship.

There is a confusion between geometric figures as objects and as symbols, and orientation is part of the difficulty. Such figures are most frequently presented in a particular orientation ('square' to the frame offered by the page), so that the orientation becomes part of the concept. Some geometric notation is iconic, e.g. □ABCD, and when the square is functioning as a symbol in this way, its orientation is important. Orientation of figures themselves is also important in that a square and a right-angled diamond are often considered by pupils to be different figures, when in fact they differ only in orientation. (See Kerslake's article, *Visual Mathematics* (1979b), for further examples.)

Repetition

Repetition as a principle is not currently widely employed as it is seen as wasteful of symbols. Usually some device, such as cipherization (the principle of using distinct symbols for different digits) or the power feature is employed in order to code the number of times a symbol is to be repeated. In the case of 88, for instance, there is repetition of the symbol, but the place-value system assigns a different meaning to each '8'. The notation xx in algebra was used some centuries ago, but x^2 is standard now.

The Ancient Egyptian hieroglyphic system did not use place value and employed a different symbol for each power of ten. A process of symbol repetition was used in conjunction with this to generate symbols for larger numbers. Because each symbol retained its relative value, the order in which the symbols were written was irrelevant. However, order was utilized conventionally, in that the symbol order for the hieroglyphs was from right to left (as with their writing generally) for increasing powers of ten. Also, the symbol for addition resembled a pair of legs walking in a certain direction, while that for subtraction was the same symbol but orientated in the opposite manner, providing another example of orientation.

Recall the earlier discussion of the undergraduate's use of ?? as a variable name. There, repetition was being used only as a device to mark a related, but different, unknown. Double and triple integrals involve symbol repetition, as do certain notations for the derivative, namely $f''(x)$ for the second derivative and $f'''(x)$ for the third. The symbol '. . .' meaning 'and so on' may look like repetition, but in fact it has now achieved the status of a mathematical logogram. Repetition has been used as an encoding principle for numerals, as can still be seen in Roman numerals, whereas, as I mentioned above, 88 is repetition in a very different sense in our decimal place-value system.

In summary, in a page of mathematical text, many if not all of the above principles will be employed in the symbolic prose. A fluent user, both reader and writer, of the mathematical writing system must be aware of these possibilities and must also be attuned to situations where they are being employed, in order to be able to interpret correctly the often subtle variations in use. In the next and final section, I turn to a deeper difficulty which can arise from attempting to make sense of symbolic mathematics.

3 Confusions between symbol and object

In Chapter 1, some examples were provided of possible confusions between the written symbols of mathematics and their referents. A teacher recalled as a pupil (one very successful in mathematics) thinking up what he thought was a very good question and saving it for the end of class to gain maximum approbation. To his surprise and consternation, this was not the effect his question provoked. His question was: 'Why do two minuses make a plus and not an equals?'. In a paper presented to an international seminar on metaphor, Reddy (1979) explores what he terms 'the conduit metaphor', which is widely employed in language about language itself. Reddy's formulation of this metaphor is that ideas are *objects*, linguistic expressions are *containers* and the meaning is *carried* by the words. An idea is 'put *into* words' or 'got across', and so the idea is conveyed by the symbol and is actually inside it. It can be asserted, for example, that 'the ideas are right there in the words'. Reddy summarises this 'powerful semantic structure in English' as 'implying that human language functions like a conduit, enabling the transfer of repertoire members from one individual to another.' (p. 311)

A couple of mathematical examples illustrate this quite well. A fairly recent innovation in the teaching of the notion of variable has been to use geometric shapes for variable symbols. A shape has the 'advantage' of acting as a container into which a number can be put, for example, $\boxed{2}$. Thus a variable is a container into which you put numbers. However, this no longer holds with literal variables, in the sense that x is *replaced* by 2 rather than them appearing at the same time on the page. A second instance is provided by the view of a set as a box. If the elements of a set are listed, e.g. $\{2,3,5\}$, then both the set and its elements are present at the same time. The more abstract notation $a \in A$ no longer has this visual spatial relation, but the relatedness is cued instead by the use of the same letter.

Reddy provides an example with reference to the text *The Old Man and the Sea*, by contrasting the following two sentences (p. 301).

The Old Man and the Sea is 112 pages long.
The Old Man and the Sea is deeply symbolic.

Such differences between form and meaning can extend even to definitions in mathematics. Here are two definitions of the concept of *evenness* which are very different in kind.

A whole number is *even*, if it ends in a 2, 4, 6, 8 or 0.
A whole number is *even*, if it can be divided exactly into two equal whole numbers.

The first definition offers a symbolic criterion, which can be used for recognizing whether or not a number is even, provided that the numbers are represented in base ten. It is not an intrinsic definition. It is inappropriate as a definition in that it does not refer to a property of numbers themselves, whereas evenness is a numerical rather than a notational property. Whole numbers *per se* do not even have digits, until they are represented in some numeration system. Evenness has to do with division, while *ending in a 4* is a surface description only. Conversely, consider the following question.

Seventeen is prime in base ten. Is it prime in base eight?

Changing the form of the representation of the number has no effect on its division properties, which relate to the number itself.

In ordinary conversation, the precise words being used are not the customary focus of attention. In mathematics, however, partly due to the unfamiliarity of the writing system, and partly due to the abstractness of all of the referents (which are mental), the symbol systems are much more commonly under scrutiny. When learning how to write, almost all of a child's attention can be taken up with the formation of the letters or the spelling of the words, which leaves little scope for *saying* anything through the writing. In mathematics, it seems that many pupils never pass this stage, and that the symbols are always uppermost as the focus of attention. They intrude and protrude, keeping the pupil's attention on the mastery of the production of the symbols themselves, rather than on trying to grasp or express the meaning which they represent.

One of the emphases of the 'New Mathematics', promulgated

in the 1960s, was making explicit to pupils the relation between symbol and idea. The particular number/numeral distinction as an overt teaching topic in the primary curriculum has been extensively criticized by many people (for instance, Feynman, 1965, and Kline, 1973). One of the reasons that the difference is unclear in mathematics is that the objects with which mathematics deals are mental constructs and hence, of necessity, abstract. In mathematics, the conventional symbol may be the only available means of evoking the concept itself.

Unfortunately, some current mathematical usage suggests the contrary, in that, for instance, the expression 'binary numbers' is widely employed, suggesting that there is a particular type or class of numbers called *binary*, rather than this being a particular means of representing numbers. Precisely the same form of expression is employed for *decimal* or *fractional* numbers, where the adjective seems to be modifying the numbers, rather than referring to a characteristic of the representation. $\frac{7}{4}$ is a fraction, 1.75 is a decimal: what is being described is the form of symbolic representation, rather than a property of the number itself. Numbers being prime or composite, on the other hand, refer to inherent properties of the numbers themselves.

'To divide by a fraction, invert and multiply' provides an example of a precept which purports to indicate how to carry out a numerical operation on fractions, namely division, in terms of an operation on the symbolic form. 'Inversion' is an operation which can be performed on a symbol where there is an inherent orientation of up and down (as well as a notation which makes use of it by breaking with the customary left–right order of writing symbols). However, the numbers represented symbolically by $\frac{a}{b}$ and $\frac{b}{a}$ are intimately connected in that they are mutual inverses with respect to the operation of multiplication.

Thus *inverse* has two meanings:

(i)   to do with 'turning upside down';
(ii) to do with obtaining 1 from a product.

It is possible however, merely by following the operations on the symbols, to obtain a correct representation of the numerical answer. Fluent users, who actually want to be able to confuse symbol and object when calculating (because it is so much quicker and more efficient), frequently forget that this mode of operation is a pathology, one potentially lethal to many learners.

One final example comes from the calculus, where the notation for the derivative $\frac{dy}{dx}$ suggests by its form that it is a fraction, while the contemporary interpretation of its meaning denies this. For many students, difficulties arise in trying to come to grips with the ideas and techniques of the calculus because of this confusion, because the metaphor *a derivative is a fraction* is not completely helpful. However, once again, there are computational advantages in sometimes treating derivatives as if they were fractions, which accounts for the continued use of the notation.

For instance, one result which is frequently expressed as $\frac{dx}{dy} = 1/\frac{dy}{dx}$ merely looks like finding the reciprocal of a fraction, yet if derivatives are algebraic fractions, why do we not just cancel the *d*s, producing $\frac{dy}{dx} = \frac{y}{x}$ (provided $d \neq 0$)?

As I indicated briefly above, the very real and frequently realized danger is that the symbols themselves, rather than the ideas and processes which they represent, will be taken as the objects of mathematics, the reality to which the language and notation is pointing and referring. Reddy describes this confusion of form with content as a 'semantic pathology', one predicated upon the conduit metaphor. Quoting Ullman (1957), Reddy (1979, p. 299) asserts: 'A semantic pathology arises "whenever two or more incompatible senses capable of figuring meaningfully in the same context develop around the same name" '. For mathematics, this confusion is so prevalent that it could easily be called *the* semantic pathology.

The 'semantic pathology' engendered by the conduit metaphor, whereby the symbol is identified with the object, involves a catachresis of the first order. Attributes of the symbols are talked about as if they were attributes of the objects themselves. 'The symbol is the object' is a powerful metaphor precisely because of the absence of certain discriminations being made. It encourages one to follow a calculation without being concerned about, or even aware of, what the objects are.

Yet, as I indicated above, there remains one very good reason for this pathology being so common. In brief, it is a useful blurring to be able to make when *doing* mathematics, operating on the symbols *as if* they were the objects being described and manipulated. Results can be obtained and calculations made by symbol manipulation, according to rules which relate to the symbols rather than dealing directly with operations on the concepts themselves. There are frequently useful surface, symbolic cues by means of which calculations can be performed, which can be codified as precepts or rules. These render facile intricate operations and manipulations in circumstances where to perform the operations at a conceptual level would require much more thought and expertise. As a colleague commented to me, a method guaranteed to produce an *incorrect* answer to a division of fractions, is to think about the meanings of the various components!

Summary

In this chapter, I identified and exemplified four different types of symbols used in the mathematical writing system: logograms, pictograms, punctuation and alphabetic symbols. In order to combine these into the various symbol clusters found in written mathematics at all levels, a number of principles of combination are used. The last section looked at a prevalent difficulty in mathematics, that of confusing symbol with object, and identified some instances where mathematicians talk about and operate on mathematical objects in ways which encourages pupils to mistake the one for the other. In the next chapter, I explore some of the higher-level organization in the structure of written mathematics and follow up this topic of the manipulation of mathematical objects by means of operations on the symbols themselves.

7
The syntax of written mathematical forms

A man a plan a canal Panama.

A fool a tool a pool loopalootaloofa.

Algebraic Graffiti

Algebra is rich in structure but weak in meaning.

René Thom, attrib.

The previous chapter detailed some of the conventional symbols employed in mathematics and how they are customarily combined to represent a range of mathematical ideas and relationships. But there is considerable structural coherence and consistency to written mathematical expressions on a larger scale which also governs their use, and, in particular, their transformation. The symbol cluster '5 + + 7 −' is as ungrammatical as a mathematical expression as *put cat the mat on the* is as an English one. Does it make sense to assign grammatical category terms such as noun, adjective or verb to particular symbols in symbol clusters? Is a different collection of functional grammatical terms such as *element*, *operator* and *relation* more appropriate to a description of mathematical syntax? To what extent is there a distinct syntax of mathematical terms and expressions, and is it consistent within itself?

1 Mathematical syntax explored

The syntactic paradigm for theoretical linguistics instigated by Chomsky in the 1950s and 1960s focused primarily on a syntax for symbolic forms, written rather than spoken. Chomsky's transformational theory of the syntax of natural language (1957, 1965; Lyons, 1984) contains two components. These are a *phrase structure grammar* and *transformations*. A phrase structure

161

grammar is a series of structural rules such as S→NP VP which indicates that a sentence (S) can be rewritten as a noun phrase (NP) followed by a verb phrase (VP). In turn, there would be *rewrite rules* (as these abstract logical schemes are known) expanding NP and VP in terms of lower-order concepts such as N(oun), V(erb) and A(djective). Finally, lexical insertion from an available set (containing actual words, a set for each of these lowest-order categories) produces the recognizable *surface* form of a sentence in the language. In this way, the notion of sentence became formalized as a grammatical construct.

Transformations operate on the output of a phrase-structure grammar (prior to lexical insertion) and are functions operating on certain strings of the symbols of the meta-language (the Ns and Vs etc.). Part of our knowledge of English includes the close similarity in meaning of sentences such as the active sentence *Myra solved the equation*, and its passive equivalent *The equation was solved by Myra*. A formulation of the passive construction, seen as derived from the active by means of a transformation, is as follows:

Passive transformation

(a) Interchange the subject and object NPs of the active sentence.
(b) Insert *by* before the new 'object'; and
(c) Insert *be* after all (other) members of the auxiliary.

(Akmajian and Heny, 1975, p. 94)

Reversing the process of derivation can assign a structural analysis to any grammatical sentence in the language, where grammaticality is determined by the phrase-structure grammar. If two different constituent structures can be assigned, which are not related by transformations, then the sentence is *ambiguous*. Commonly-cited examples include the sentences, 'Flying planes can be dangerous' and 'The police were ordered to stop drinking after midnight.'

A working programme for mathematics, modelled on the above analysis, might therefore be to attempt to derive a phrase-structure grammar for parts of mathematics (two obvious contenders being arithmetic and algebra) and then to attempt a transformational account operating on the output. One question is what is to be the meta-language for describing transformations in both of these cases.

Here are some examples of a transformational account of certain operations from arithmetic. There is ambiguity in the mathematical 'phrase' $3 - 4 - 5$, in that it can be bracketed as $3 - (4 - 5)$ or $(3 - 4) - 5$ which have different meanings. For example, calculators with different internal logics will evaluate the string $3 - 4 - 5$ differently. There are two different possible structures.

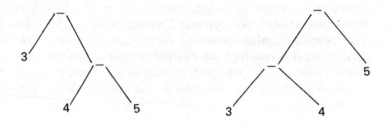

$3 - 4 + 5$ has a similar ambiguity, whereas $3 + 4 + 5$ does not, which suggests a transformation taking

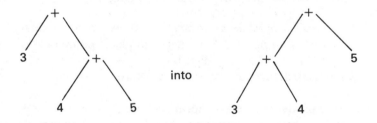

and conversely. This is recognizable as being equivalent to the statement that the operation '+' is associative.

What is the meta-language to be employed to express this transformation, presuming this to be a generic instance? An obvious candidate seems to be abstract algebra, and 'generalised arithmetic' is one of the common approaches to introducing algebra in schools.

Thus an (optional) transformational rule a∗(b∗c) → (a∗b)∗c (and its converse) together with a marker against the operation of '+' in the (mathematical) dictionary will account for the (mathematical) synonymy of 3 + (4 + 5) and (3 + 4) + 5, and hence the absence of ambiguity in the phrase 3 + 4 + 5. In the above, I have been implicitly using the definition that an arithmetical phrase is a finite, grammatical string of numerals and operation symbols, where phrase synonymy is numerical equality. (I am aware that *grammatical* is a question-begging term, in that I have not yet proposed a phrase-structure grammar for arithmetic. Among other things, such a grammar would have to rule out the example given at the beginning, namely 5 + + 7 −.)

Below are a couple of arithmetic transformations, formulated in the style of linguistic ones (e.g. Akmajian and Heny, 1975). Basically, what is required is a description of the general form of an acceptable input (*structural description*), together with a description of the change effected by the transformation (*structural change*).

Reducing (unsigned) Fractions
e.g. reduce $\frac{16}{24}$ to lowest terms. $\frac{16}{24} \to \frac{8}{12} \to \frac{4}{6} \to \frac{2}{3}$
Structural description: $\frac{p}{q}$ with p, q whole numbers, $q \neq 0$.
Structural change: $\frac{p}{q} \to \frac{r}{s}$ where r, s are whole numbers with $p \times s = r \times q$ and also with $0 \leqslant r < p$; $0 < s < q$.
This describes any correct fraction reduction.

Extended place value notation
Structural description: Any integer in place-value notation i.e. $x_n x_{n-1} \ldots x_1$ each x_i a digit 0–9.
Structural change: $x_n \ldots x_1 \to x_n \times 10^{n-1} + \ldots + x_1$

This rule produces a surface form such as $23 \to 2 \times 10 + 3$. Just as with linguistic transformations, the structural descriptions are in terms of generalized surface strings, which are seen as linear collections of concatenated symbols. A few general points can be made about these examples, some of which contrast with natural language transformations.

1   The meta-language of arithmetic is algebra. Many, but not all, of the transformations – the laws of arithmetic – are either taught explicitly to the pupils in the meta-language, or

in the form of arithmetic specializations from which the pupils are expected to extrapolate the general form. There remains the ambiguity of whether these are the laws of arithmetic or the laws of algebra.

2 Transformations are often reversible, but are frequently applied only in a left-to-right manner and mutually-inverse pairs are perceived and described differently. For instance, if the notion of phrase is extended to include letters, thereby switching from the context of arithmetic to that of algebra itself, then:

Expanding brackets:

$(x + a)(x + b) \rightarrow x^2 + ax + xb + ab$ (This is now the *surface* form in algebra.)

and

Factoring:

$x^2 + (a + b)x + ab \rightarrow (x + a)(x + b)$

are seen as two separate actions, and hence, transformations.

3 The prevalent language of the teacher in describing such transformations is often purely in terms of the surface structure, thereby focusing attention on the form rather than the meaning which gives rise to the transformation.

Take it over to the other side and change the sign.

Cross multiply.

Move the decimal point over.

Turn it upside down and multiply.

Collect all the xs on one side of the equation.

Always do to the top what you do to the bottom.

To multiply by ten add a nought.

All of these descriptions are to do with actions on the symbols, on how to achieve grammatical utterances from other ones. They tell what should be done, with the result that there is little impetus to examine them to see *why* they might be helpful transformations to carry out. Mathematically interesting things can occur if they are *not* blindly followed on every occasion. See Brown and Walter (1983) for an exploration of one outcome of ignoring the injunction to 'Collect all the xs on one side of the equation.'

4 Mathematical syntactic rules are prescriptive and are consciously and deliberately taught, learned and applied at the

surface level. When confronted with a student who had written $\int e^{dx}$, I heard myself say 'You never have a dx as an exponent.' I was then asked why such things had not been taught to him before. My presumption that the meaning guided the correct choice seemed somewhat naive in the face of students for whom mathematics was the formulation and manipulation of strings of symbols according to various rules.

All the transformations I have been discussing up until now have been phrase-to-phrase ones, which I call *structural*. Every computational algorithm (e.g. to evaluate 27×36 or $349 \div 17$) can be seen as a finite, *ordered* sequence of structural transformations. More generally, there are partial orderings among the applications of certain transformations, as evidenced for instance by the acronym BODMAS (Brackets, Of, Division, Multiplication, Addition, Subtraction).

Two different algorithms for subtraction are given below and to comprehend why they work requires an awareness of quite intricate transformations.

Consider the subtraction:

$$\begin{array}{r} 52 \\ -37 \\ \hline \end{array}$$

(a) You 'borrow' 10 and change the 2 to a 12 and 'repay' it (note bourgeois morality at work in the terminology) by changing the 3 to a 4. In other words, this algorithm solves $62 - 47$ (or, more accurately, fifty-twelve minus forty-seven), using the equivalence
$a - c = (a + k) - (c + k)$ in a special case.

(b) Using expanded notation is a way of rendering less opaque the compact place-value notation of our decimal numeration system.

$$\begin{array}{rcccc} 52 & = & 50 + 2 & = & 40 + 12 \\ - \ 37 & = & 30 + 7 & = & \underline{30 + \ \ 7} \\ & & & = & 10 + \ \ 5 = 15 \end{array}$$

i.e. $(a + b) - (c + d) = (a - c) + (b - d)$

If the algebraic phrases themselves are the focus of attention, as in the examples of expanding brackets and factoring, the meta-language itself becomes the object of study, and algebraic transformations, the 'laws' of algebra, are themselves formulated algebraically. Algebra is its own meta-language. Unfortunately, there are instances where the form of certain algebraic transformations conflicts with particular arithmetic ones, providing further interpretative difficulties.

One example is × deletion. The simplest example of this transformation is $a \times b \to ab$. This is a surface-tidying operation, possibly also to remove the real possibility of confusion between the letter x and the multiplication symbol. Care, however, has to be taken if subsequent arithmetic specialization occurs, in that if $a = 2$ and $b = 3$ (an assignment reminiscent of the insertion of lexical items), then $ab = 6$ and not 23, which is a common error of pupils learning algebra. Another related difficulty arising from confusing the level and formulation of the transformation comes from multiplication being commutative $(a \times b = b \times a)$. When combined with the arithmetic transformations '× deletion' and 'expanded notation', this transformation produces:

$$a \times b = ab = 10a + b$$
$$b \times a = ba = 10b + a$$

Yet $10b + a \neq 10a + b$ in general.

Juxtaposition of symbols has different meanings in arithmetic and algebra. This would seem to be clear evidence for a rule of '+' deletion, arising from e.g. $2 + \frac{1}{2} \to 2\frac{1}{2}$, $2 + 0.5 \to 2.5$ and also place value, $20 + 5 \to 25$. The phenomenon of over-generalization of the applicability of rules may account for many pupils applying + deletion at the level of algebraic transformations to render, for instance, $3a + 5b \to 8ab$. (See Hart, 1981.)

In general, possibly as a result of prompts to algorithmic transformations (merely writing $6 + 9$ can be perceived as a powerful prompt to carry out the addition), there seems to be a pressure to apply transformations until all of the operation symbols (verbs?) have been removed by performing them. I earlier cited Booth (1984) who describes pupils who calculate $j - 3$ as g, the alphabet providing a context in which the subtraction could be meaningfully carried out. This would also result in the observed unwillingness to accept $2a + b$, or even $\sqrt{2}$, as an acceptable final surface string of symbols.

Pupils often have internalized incorrect, but plausible, transformations e.g. $(a + b)^2 = a^2 + b^2$ (possibly seen as an application of the precept 'do the same thing to both sides') or $\sqrt{a + b} = \sqrt{a} + \sqrt{b}$. They are incorrect in the sense that they are not unconditional transformations (identities), whereas $(a + b)^2 = a^2 + 2ab + b^2$ is. One approach for contending with such 'mistakes', which seem highly resistant to change (certainly one counter-example is insufficient to dissuade), is to focus explicit attention on the transformation and ask and explore questions about its scope and validity. In this way, the pupil is directly attacking the question of where transformations come from, and, one might hope, seeing that transformations are comprehensible as expressions of high-level generalizations. Such questions might include:

1) When are they valid within the existing system?
2) Can we change our system, possibly relinquishing other transformations, so that this one *will* be valid?

A changed system will permit different transformations, and so more sense might be made of the switch from not being able to transform $4 - 8$ ('eight from four, you can't') or to solve $x^2 + 1 = 0$, to being able to, as a consequence of the introduction of negative and complex numbers respectively.

There is a separate class of mathematical transformations which operates on sentences as a whole. For the purposes of this section, I will take a *mathematical sentence* to be a pair of phrases linked either by an '$=$' or one of $\geqslant$, $>$, $\leqslant$ or $<$, and the sentence is called an *equality* or *equation*, or an *inequality respectively*.

eg. $217 + 203 = 420$
   $2x + y > 16$
   $ax^2 + bx + c = 10$

As I mentioned earlier, transformational grammar applied to natural language attempts to account for some of the relations we perceive among sentences. In mathematics, the sentences $7 - x = 4$ and $4 + x = 7$ are similarly perceived as related (synonymous). A first attempt at describing mathematical sentential transformations might be as a sentence-to-sentence mapping where the relation of synonymy becomes 'has the same solutions as'. Unfortunately, there are some transformations such as squaring both sides, which can introduce spurious solutions, while retaining the genuine ones. Hence there is a need for a weaker formulation, namely, 'the solutions of the output sentence contain those of the input sentence'.

The following sentential transformation, unless adequately formulated, when mis-applied can result in the *loss* of solutions.

Cancellation

Structural description: $P_1 \times P_2 = P_1 \times P_3$
Structural change: $P_2 = P_3$ (Condition $P_1 \neq 0$)
e.g. $x^2 + x = 0$ 'Divide both sides by x' (an algebraic
                                                 formulation)
     $x + 1 = 0$ 'add -1 to both sides'
     so $x = -1$.
The solution $x = 0$ is 'lost' as a result of using cancellation as a universal transformation.

The introduction of the '$=$' as an essential component of a mathematical sentence often causes difficulties, as it has also previously been used as a link between a phrase and its transform. The relationship between a sentence and its successor in a mathematical argument is apparently unclear to many students, and a common strategy is to link every phrase by an '$=$'. Thus, the problem,

If $f(x) = x^2$, find $f'(2)$

provoked one student to write:

$$f(x) = x^2 = f'(x) = 2x = 2 \times 2 = 4$$

In this section, I have attempted to describe some of the transformations, both structural and sentential, which can be found in arithmetic and algebra. The mathematical equivalent of the NP VP meta-language for natural language (in which the transformations are formulated) is algebra in both cases. We expect pupils to be fluent users of this meta-language, as well as being able to distinguish whether a transformation is applicable to the world of algebraic 'objects' or to that of arithmetic ones. In the next section, I shall turn to recent work in computing which has utilized such a formal structural approach in describing the transformations which mathematicians perform on symbols.

2 Modelling approaches to mathematical transformations

Approaches to natural language using a computer have been extensively reviewed by Ritchie (1982). In mathematics, particularly given the string manipulation facility of the LISP family of computer languages, varied attempts have been made to get computers both to perform symbolic (as opposed to arithmetic) calculations and to prove theorems. (See Bundy, 1984 for a descriptive summary.) In the context of mathematics education, there are three programs I wish to discuss.

The first two are very similar in form: Subtraction Buggy (written in SMALLTALK) developed by Seely Brown, and Fraction Buggy (written in PASCAL) developed by O'Shea and du Boulay. They both attempt to model faulty child procedures for performing the eponymous mathematical operations. In the introduction to the fraction program, the authors claim,

> Children often make mistakes when doing fraction sums and sometimes these mistakes are just slips. But very often mistakes are due to misunderstandings about fractions which cause the child to use the same procedure consistently . . . This program calls these incorrect, but consistently applied, procedures 'bugs'.

Fraction Buggy is used as part of the Open University course EM235 *Developing Mathematical Thinking* to encourage teachers to notice and explore the more systematic aspects of pupils' errors. The program operates by responding to fraction questions (with an open choice of numbers and operations) posed by the user, just as if an actual child had been asked to do the problem. An example of a generated data set is as follows.

$$1\tfrac{3}{4} - \tfrac{1}{2} = 2$$
$$\tfrac{2}{3} - \tfrac{1}{3} = 1$$
$$\tfrac{7}{8} - \tfrac{5}{8} = 2$$
$$\tfrac{8}{9} + \tfrac{1}{3} = \tfrac{1}{6}$$
$$\tfrac{3}{7} \times \tfrac{7}{8} = \tfrac{3}{8}$$
$$\tfrac{1}{8} \div \tfrac{1}{4} = \tfrac{1}{2}$$

If you believe you have diagnosed the bug(s), the program then probes your conjectures by sample questions, asks for your predictions, and then responds with what would have been so-and-so's response (each different bugged algorithm is given a person's name). However, assuming no random component is programmed in, the machine applies the rules completely consistently, which facilitates diagnosis in this simulated context. Nonetheless, even in this protected environment, diagnosis is by no means a simple task.

It is an extremely interesting question concerning how systematic and procedure-mimicable children's errors are, and also whether this approach will prove useful for teachers. Discussion of such an approach to errors can be found in Ginsburg (1977) and also in a number of issues of the *Journal of Children's Mathematical Behaviour*. (See e.g. 1978 (*1*) or 1979 (*2*).) Although the opening discussion of the simulation suggested the metaphor of flawed understanding giving rise to 'bugged' procedures, possible processes of debugging are not explored specifically in the program, although they are in the course EM235 itself.

The second example I wish to discuss reflects an approach to solving equations, which attempts to represent codified human knowledge about this activity in the program's procedures. PRESS, an acronym of PROLOG Equation Solving System), developed by Bundy (1984), is designed to emulate the written protocols of expert equation solvers. PRESS performs a series of manipulative transformations on symbols, guided by a set of contextually-based principles to monitor the solution process and to ascertain what might be a good next operation to perform. The latter is essential in order to prevent a 'combinatorial explosion' arising from the exponential growth in the number of possibilities. This is because a considerable number of transformations have structural descriptions which are fulfilled by many outputs: for example, $u \times v \to v \times u$ (where u and v are strings of symbols) or $(u + v)(u - v) \to u^2 - v^2$ (and its converse).

One of the examples Bundy uses to discuss the program's operation is the A-Level question:

Solve $\log_e(x + 1) + \log_e(x - 1) = 3$

The machine's solution proceeds as follows:

$\log_e(x + 1) + \log_e(x - 1) = 3$

$$\log_e(x + 1)(x - 1) = 3$$
$$\log_e(x^2 - 1) = 3$$
$$x^2 - 1 = e^3$$
$$x^2 = e^3 + 1$$
$$x = \pm \sqrt{(1 + e^3)}$$

In obtaining this solution, there are a number of processes at work. By using tree diagrams, the machine is able to identify precisely where the occurrences of the unknown x are, and some of the general aspects of the program are to ensure that:

a) the 'distance' between the various occurrences of x is decreased (*attraction*);
b) terms are combined so as to produce a single occurrence of the unknown (*collection*);
c) then the program strips away (via inverse functions) the expressions to obtain x, decreasing its depth of nesting at each stage (*isolation*).

These are meta-level concepts providing syntactic information about the way algebraic equations are represented in the machine, just as it is on the surface properties of the representation of the equation that the computer operates. I should point out that the above is merely a superficial summary of some of the design features of a very sophisticated program. PRESS is very successful at solving A-level equation tasks, and it seems clear that, although at the moment the program requires a large amount of space in the computer (as does Fraction Buggy), within a very few years there could be programmable calculators, fitted with PRESS-tell chips, readily available in schools.

Computers are very successful mimics, particularly where a clear decision algorithm is being employed. A measure of this program's achievement is that for equation solving, the situation is nowhere near so clear-cut. PRESS is more credible as a model of mathematical thinking than are natural language programs for natural languages. This is because it is implementing the consciously formulated and articulated rules, procedures and principles of expert human solvers. Such a general design procedure results in what has come to be called an expert system. O'Shea and Self (1983, p. 42) describe these as 'programs which try to solve problems which would otherwise require specialist human skill.'

However, I have noticed a tendency to presume that because there is a program which can mimic the surface features of some activity, there has therefore been provided:

a) deep insight into what is necessarily required to do that activity,
b) (even more tendentiously) insight into how humans do it.

Three questions come to mind.

1) Does a concentration on the manipulative aspect of mathematics necessarily provide insight into the pre-requisites for that activity?
2) Does the fact that a human can program a machine to perform manipulations in a way that 'saves the pheno-mena', necessarily indicate that all (any?) humans would use the same approach?
3) Even if the answers to both of these questions were unequivocally yes, would this help us to teach equation-solving to human novices?

Bundy has commented that if pupils understood the nested, embedded nature of variables such as x in complex compound expressions such as $\log(\sin(2x^3 + 1))$, rather than perceiving them as merely surface, linear strings of concatenated symbols, certain widespread symbolic errors would disappear. This claim is based on a move from knowledge of the program (in particular, the meta-concepts employed to represent algebraic knowledge) to suggestions for remedial teaching. It seems to me that, just because the program needs to 'know' this explicitly, it does not necessarily imply that pupils will need to at a conscious level. It is not clear to me that overt instruction about structural knowledge, particularly of detailed technical matters of this sort, is at all what is required.

Let us look at the conception of algebra involved. Recall the discussion of what constitutes algebra in Chapter 5 and in particular the Hewitt quotation on p. 136. Algebra, *in terms of this program*, is nothing more than the application of rewrite rules to expressions. Within this framework, a successful mimicry of surface mathematical behaviour is entirely possible. Such a perspective is remarkably close to a formalist view of mathematics as a whole, namely that mathematics is the rule-governed manipulation of marks on paper. To this extent, computer modelling can be seen as the latest refuge of the behaviourist.

Sadly, I fear, this description of algebra in particular, and mathematics in general, would find a strong resonance with the experience of many pupils in mathematics classes. Humans seem, in the main, less adept at rule learning and application than machines, *in the absence of meaning*. Because meaning is largely absent for many people in mathematics, they are forced back on to attempting to learn the features *directly* for generating correct grammatical expressions. That is, they are trying to learn an apparently arbitrary syntax for symbolic forms. The surface features in mathematics are more often consciously taught: for instance, 'there will always be a '*dx*' at the end of an integral' or 'a differential equation is an expression of the form . . .'. Yet in the absence of understanding, little difference will be perceived between, for example, $\int_1^2 \frac{1}{x} dx$ and $\int_{-1}^1 \frac{1}{x} dx$, as both have the correct syntactic 'features' for definite integrals.

Pupils who, in response to the injunction to 'collect all the *x*s on one side of the equation', perform precisely that and transform, for example, $x + 4 = 3x$ into $xx = 34+$, are acting to some extent in keeping with this computational tradition of employing the surface, symbolic features as the focus of awareness and basis for action. Too much algebra teaching is solely syntactic, in that much mathematical practice is coded into precepts which operate entirely on the symbols, rather than being combined with a meaning (and hence a purposeful goal), an interpretation in which the requisite transformations make some sense.

I wholeheartedly agree with A. N. Whitehead when he claims that, 'Civilisation advances by extending the number of important operations we can perform *without* thinking about them' (1925, p. 59, my emphasis). To this extent, calculators and computers are liberating devices, freeing the mind of the technical intricacies of computational algorithms, thereby permitting the addressing of more serious strategic problems, such as the choice of operation appropriate to a particular problem. An equation-solving chip could be an equally powerful and enabling tool, providing an entry to the use of techniques which were formerly inaccessible.

However, we should not permit the operational efficacy of

powerfully dissembling software to distort or deny the purpose and reason which mathematicians see in their activities. 'Every mathematician endowed with any intellectual honesty will agree that in each of his proofs, he is capable of attaching a meaning to each of the symbols he manipulates' (Thom, 1971, p. 696). The success of PRESS on a diet of A-level equation-solving problems, testifies, in part, to the vapid nature of those tasks, rather than the successful modelling of mathematical activity itself. Similarly, giving a calculator to an O-level or particularly CSE examination candidate will frequently render trivial a substantial proportion of the questions.

Notice the word *attaching* that Thom used. It is generally accepted among mathematicians that for complex calculations, which have become routine, it is often easier to 'disengage' the semantic component, and operate formally, by rule, on the symbols alone. School teaching at all levels seems to have accepted the goal of symbolic algorithmic fluency, without sufficient concern for semantic re-integration. One way of expressing this moral is that it is far easier to *do* A-level mathematics, for example, than to understand it.

In the introduction to his book, *Evolution of Mathematical Concepts*, mathematician R. L. Wilder makes an enlightening distinction between two possible ways of reacting to and dealing with symbols. The first is (my emphasis):

> Man possesses what we might call *symbolic initiative*; that is, he can assign symbols to stand for objects or ideas, set up relationships between them and operate with them on a conceptual level.

Wilder goes on to discuss a contrasting (and regretfully more common – at least in mathematics) mode of operation:

> . . . what we might call *symbolic reflex behaviour* . . . (they) do not create the symbols, but can react to them just as they react to other environmental stimulants. However, much of our mathematical behaviour, which was originally of the symbolic initiative type, drops to the symbolic reflex level. We memorise multiplication tables and then learn special devices (called algorithms) for multiplying and dividing numbers. We memorise simple rules for operating with fractions and

formulae for solving equations. These are justifiable labour-saving devices and the professional mathematician understands the purpose of what he is doing, while the pupil, who only learns the devices, usually does not even comprehend why they work.

It is therefore personally advantageous to have as many mathematical operations as possible at the symbolic reflex level, to ease the cognitive load. But this is only so *provided*, and this is crucial, that you can, at will, reinstate such 'reflexes' as objects of understanding and subject them to conscious thought and control. It is really a question of power and mastery in relation to the use of symbols; whether you use them and can make them work for you, or the converse. Wilder concludes:

A considerable amount of what passes for 'good' teaching in mathematics has become of the symbolic reflex type, involving no use of symbolic initiative . . . What essential difference is there between teaching a human animal to use an algorithm to find the square root of a number, and teaching a pigeon to punch certain combinations of coloured buttons that will produce food? (Wilder, 1968, p. 3)

The notion of symbolic initiative runs counter to the prevailing emphasis on the strict conventional use of certain classes of symbols (e.g. single letters as variables rather than question marks) and even particular letters in certain instances. Around the turn of the century, when geometry teaching was still firmly subordinated to the teaching of Euclid, great emphasis was laid on being able to reproduce exactly the figures and constructions of the various propositions. This desire for faithful reproduction also extended on occasion even to the letters denoting the vertices of the figures, and a response was in danger of being marked wrong were these particular reference symbols not strictly adhered to. Such precision in sticking to a stereotype of 'good' mathematical practice ceases to have anything to do with mathematics whatsoever, and crucially misunderstands the role of symbols.

In conclusion

In this chapter I have tried to explore one interpretation of the notion of the grammar of natural language in the context of arithmetic and algebra. Transformational grammar proposes a meta-language and a set of concepts with which to describe formally some of the transformations which seem to be present in natural language. It also provides a generative scheme by which to assign constituent structures. I use the term 'seem' merely to suggest the question of whether or not the rules, of which some linguists are seeking a precise formulation, have psychological reality.

This claim for the psychological reality of such structures is frequently made, despite many native speakers denying any conscious access to or knowledge of explicit rules governing the formation of sentences. Virtually all four-year-olds are grammatically competent in languages whose syntactic structure adult linguists are unable to come close to describing as a whole. As an analogous claim, it could be argued that *because* the music of Mozart, which he composed at age five, exhibits classical harmony, he *therefore* must have had knowledge of or understood the content of such a theory.

(To have knowledge of something need not, however, imply that it is consciously accessible or alterable. It makes sense to claim we have knowledge of how to regulate our own breathing and temperature, as well as how to digest our food. Yogis have indicated how, with remarkable effort, it is possible not only to become deeply aware of such processes but also to begin to control them on a conscious level by an act of will. At a more mundane level, many have knowledge of how to ride a bicycle, yet find it extremely difficult to articulate at all precisely or satisfactorily what it is or even how it is known. How could someone argue with a formal theory of what such tacit knowledge consisted?)

Some of the consequences of such an approach may be seen, in part, by the powerful computer programs I have discussed, as well as more general symbol manipulation packages which are just becoming available as I write (such as BOX, MAPLE and REDUCE). These may very shortly become the mathematical analogue of the more familiar word-processing packages. Whole programs such as PRESS might shortly become routine components of such systems for manipulating mathematical symbols according to prescribed algorithms.

However, although such an approach affords a succinct description of surface operations on symbols, and allows computer programs to be written which, for example, mimic equation solving, it also encourages such a description to be used as a vehicle for teaching. With apologies to Kant, while semantics without syntax is blind, syntax without semantics is empty. Mathematics is not the manipulation of symbols according to prescribed rules: mathematical activity can be both purposeful and meaningful to human beings.

I should like to end this chapter with a personal anecdote. I had trouble understanding chemistry at school because, among other things, I treated chemical equations as algebra. I did so because that was how I saw them, letters, numbers, operation symbols and transformations on them. I could *do* algebra. I learnt with little difficulty the rules about making such *chemical* equations 'balance'. I failed, however, to understand why some of my equations were acceptable and others were not. If I had an equation which balanced algebraically, I was unable to see why it was wrong – because the *form* was correct. I now make sense of this experience in part as an instance of a symbolic register confusion. What I was missing was any sense that chemical equations were describing certain aspects of the world (some parts of which steadfastly refused to combine with others, despite my symbolic permission and approval), rather than merely being manipulations within a closed symbolic world according to specified rules.

Structural analysis and powerful programs are very important and impressive, but it is essential to retain a sense of proportion and direction. Our society has seen an increasing incursion of machines from the automation of the physical to that of the mental. Machines can extend our reach and strength, but can also restrict and atrophy our current abilities, as well as distort our perceptions of what we ourselves do and why.

8

Reading, writing and meta-linguistics

Take care of the sense, and the sounds will take care of themselves.

Lewis Carroll, *Alice's Adventures in Wonderland*

There is a very large quantity of written mathematics, at widely differing levels of complexity. In order to gain access to this stored information, pupils need to know how to read mathematics. In the two previous chapters, I have indicated how complex the mathematical writing system is. What relation does written mathematics have to spoken natural languages such as English which are equipped with a mathematical register? How are the structural metaphors discussed in Chapter 4 reflected in the notations of mathematics? These two themes provide the core of this chapter which attempts to re-integrate the written and spoken strands which make up this book by means of the notion of reading. I then conclude with a brief examination of the recently burgeoning field of meta-linguistics (the study of the knowledge and beliefs of individuals *about* language itself) and explore its relationship with mathematics.

1 Reading mathematical text

There are at least two major senses in which the word *reading* can be used. The first refers to an ability to produce a sequence of sounds in a language when faced with a written text, commonly called 'decoding to sound'. (It has also, more graphically perhaps, been termed 'barking at print'.) The second sense of *reading* requires the abilities implicit in the first sense together with the requirement of 'understanding' what is being read. I do not wish here to get into an attempt at a detailed specification of what this latter term entails, but at a simple-minded level, I hope the distinction is clear.

179

In the former interpretation, reading is a form of translation ability enabling the reader to alter language from one realization to another, a counterpart of dictation. No mention, however, is made of understanding. Thus, I can be said (according to this view) to read fluently in German in that I can pronounce almost all the words in such a way that a native German speaker could understand them, quite irrespective of the fact that I will probably understand very little of the meaning of what I read. I do not know German, but I can sound as if I do. How is a German speaker to know that I cannot understand what I can read so fluently? In part, this question harks back to the brief mention of the ELIZA program in Chapter 1 and the fact that native speakers seldom separate out various language abilities and presume global competence on the evidence of partial facility.

Even with this lesser conception of reading, there is a difficulty in relation to reading written mathematics. In a *verbal* or *mixed* mathematical text, there will be words from the mathematics register as well as that of 'ordinary English'. While many of the technical terms of the former have Latin or Greek origins, they conform to the standard grapheme– or morpheme–phoneme relationship (the links between clusters of written symbols and spoken sounds which form the basis of reading many natural languages) and can be learnt to be read much as any new word. (See Stubbs (1980) for further details about English.) Thus *dodecahedron* or *parallelogram* can be read in the same way that words like *antidisestablishmentarianism* can.

However, with *mixed* or *symbolic* mathematical texts, there is the more complex question of how to read the various symbols which form the mathematical writing system. In Chapter 6, I classified the individual symbols into four classes: logograms, pictograms, punctuation symbols and alphabetic symbols. For logograms and pictograms, there is only one basic possibility, namely to read it as a linked English word or phrase that has been learnt. Thus $2 + 3$ can be read variously as 'two plus three', 'three added to two', 'two and three' and so on. If we wish to refer specifically to the symbol '$+$', it is usual to call it 'the *plus* sign'. In order to *read* it you need to know its name(s). If you do not know it, you cannot *spell* it out (as the mathematical writing system is not alphabetic in nature). The written representation gives no help, nor is there anything like a one-to-one correspondence between signs and possible vocalizations.

For punctuation and alphabetic symbols, there are two general possibilities: reading at the level of meaning and reading at the level of the symbol *qua* punctuation or alphabetic symbol. For example, the symbol cluster $\frac{dy}{dx}$ can be read at the level of symbols as 'd y over d x' and at the level of meaning as 'the derivative of y with respect to x'. *Both* are common and accepted articulations. In the latter case, the meaning has been used to guide the vocalization; in the former, it was a symbol-by-symbol pronunciation, linked by a description of the symbols' relative positions. (This too, however, is partially conventional, in that 'dx under dy' is not an accepted reading.) To be able to read such symbols correctly requires access to the various structurings described in Chapter 6.

Let me provide a couple of further examples. 27_{FIVE} is a notation found in school texts on multi-base arithmetic. It can be read at the level of symbols (which I shall henceforth call a *spelling pronunciation*) as 'two seven subscript five'. It can be read in the second sense (henceforth termed *interpretative*) as 'twenty-seven base five'. The interpretative reading requires knowledge of both place-value notation as well as the positional subscript notation, where the subscript number is written in its verbal form. As a second example, the notation a:b has a spelling pronunciation of '*a* colon *b*' and an interpretative reading of 'the ratio of a to b'. In both of these examples, the interpretative reading is the usual one. It can already be seen how with reading mathematical symbolism, the two senses of reading I distinguished at the outset are coalescing somewhat.

Since every punctuation device and alphabetic letter has a word or sound associated with it (e.g. *alpha, comma, el*), almost all mathematical statements have a spelling pronunciation. As I mentioned earlier, in the case of logograms, they need an associated term or to be described as a geometric shape. In the case of the digits 0–9, the name for the symbol is the same as the English word for the concept. 'What is this?' 'It is a two.' Open University students (who study at home and seldom hear mathematics spoken) regularly report difficulties in not being able to articulate to themselves (whether vocally or subvocally) as they are reading their printed units. This is precisely because they do not know how to *read* many of the individual symbols (e.g. unfamiliar letters or logograms), nor do they have access to the interpretative readings of the symbol clusters.

As a sixth-form pupil, I had a problem of reading in the context of matrices. How should the first element in a matrix a_{11} be rendered into spoken English? Operating in terms of familiar components, '*a* eleven' or '*a* subscript eleven' would be a reasonable attempt (the first reflects the order of the components, the second both the order and the relative position). Notice how important in mathematics the notion of the main line is in providing a reference basis from which to work. Both of these readings are part interpretative and part spelling. Unfortunately, the actual pronunciation is related to function, whereby the subscript represents the element's location in a two-dimensional array, and is read '*a* one one'.

Ironically, this sounds like a reversion to a spelling pronunciation of 11, but actually reflects the fact that there are two numbers rather than just one. Stubbs (1980, p. 6) remarks, 'any writing system has to compromise between the requirements of writers and readers.' The compactness (and hence speed) for the writer of a_{11} rather than other alternatives seems to have overriden reading or novice considerations. Possible variant notations for the same idea include $a_{(1,1)}$ (as in Cartesian coordinate locations), $a(1,1)$, which emphasizes the function aspect from a domain of *nm* integer pairs to the real numbers, or even a compromise notation $a_{1,1}$ (where the comma in the surface form indicates that there are in fact two different numbers present). This provides an instance to support the claim that the mathematical language is designed for experts rather than novices. In some cases, such as the above example, the notation is sub-minimal for reading, in that not all the required distinctions for reading are present in the written form.

Interpretative reading versus a spelling pronunciation can, on occasion act as a guide to the teacher about comprehension. There are other examples which may or may not reflect misconceptions, as they are the commonly accepted readings among mathematicians. The numeral 0 is often read as 'oh' (as if it were the letter) rather than 'nought' or 'zero'. The spelling pronunciation is actually of a very similar symbol rather than of the logogram 0 itself. The inverse (under composition) of a function *f* is denoted f^{-1} and read '*f* minus one' or '*f* to the minus one'. The inverse (under multiplication) of a number *a* is written a^{-1} and read as '*a* to the minus one' and equals $\frac{1}{a}$. Unfortunately,

$f^{-1} \neq \frac{1}{f}$, yet the identical spelling pronunciation of these identical forms can suggest this.

The accepted readings of mathematicians tend to be a mixture of spelling and interpretative pronunciations. Spelling pronunciations are used for the very good reason that, unlike in English, 'spellings' tend to be much shorter. Consider the symbolic phrase $\Sigma_{i=1}^{i=n}i^2$. 'Sigma one to n eye squared' is considerably shorter than 'The sum of the squares of the first n natural numbers' or 'the sum of the squares of the natural numbers from one to n.' Consider $\int_0^1 \sin(x)\, dx$. 'Integral nought to one sine x dx' is shorter to say than 'the definite integral from nought to one of the function sine of x with respect to x'. Notice with the logograms $\int$, 0 and 1, there are no 'spellings' available. Once again, mathematicians who can detach the semantic component of the symbols at will can work much more quickly with the symbols. Once again, for the novice, it *sounds* as if the symbols are the objects of mathematics. The habits of the initiates are not necessarily best suited to the needs of the novitiate.

There are many more issues in reading and mathematics. In the area of textbooks some work has been done on the question of readability (see Kane *et al.*, 1974, and Shuard and Rothery, 1984). The notion of a measure of text readability independent of the reader seems to me to be misguided, and stems from a theoretical position with regard to the semantics of reading (i.e. the second conception of reading I delineated at the outset). This is that the meaning inheres in the text itself, and therefore it is possible to measure how difficult it is to extract it in an objective way. This mining metaphor has been roundly criticized by Wing (1985) in a review of the Shuard and Rothery book.

There is also the question of the structure and syntax of the mathematical writing system and how it relates to that of the articulating language. Mathematical symbolisms have, in the main, emerged from mathematicians writing in Latin and then European languages, so there is no likelihood of a major mismatch for these languages. What about languages which have widely different syntactic structures? (See UNESCO, 1975.) One aspect is that of word order. There are certainly a few differences even in English. For instance, the numeral 18 is read as 'eighteen', where the individual morphemes do not have the same order as their symbolic counterparts. 'Onety-eight' would be a reading more consistent with the general structure of the symbolic decimal numeration system. In addition, although in general the left-to-right order is preserved, the fact that many mathematical symbols make use of vertical position (that is, they are two-dimensional in nature), requires some adaptation of this. This comes in part from having to map a two-dimensional configuration onto a one-dimensional (time being linear) flow of sound.

Finally, I would like to mention the problem of how mathematicians write for each other. Reading mathematical writing is extremely difficult due in part to the lack of redundancy in the writing system and partly to the prevailing values of professional mathematical writing. Elegance is measured in part by brevity and in part by simplicity. Accessibility plays no part. Because of structural differences between mathematical and English prose text, a different style of reading needs to be adopted by the reader, and pupils need considerable training on *how* to read mathematics.

2 The written mathematical register: notational metaphor

Although it is customary to refer to registers in relation to spoken language, it also makes sense to think of the written register of mathematical symbolism: that is, the collection of symbols and the conventions governing their combination and use to convey mathematical meanings. I discussed in Chapter 5 some examples of what can be seen as register confusion, with regard to the algebraic and non-algebraic use of single letters. I also wish here to pick up the theme of structural metaphor and look briefly at its reflection in the symbolic notations of mathematics. It will also tie in with the previous remarks on reading, in that interpretative reading requires comprehension of the symbols, and if a symbol is used metaphorically it may be read differently according to the various uses. For example, the same multiplication symbol is read as 'times' in a whole number context and 'of' in a fraction context, and possibly not at all in an algebraic one.

My first example of a notational metaphor comes from the domain of arithmetic. Contrast the statement $3 - 2 = 1$ with $2 - 3 = -1$. The latter expression involves an unexamined metaphor which leads us to presume that the meanings of certain symbols (for instance, 2, $-$, $=$) are the same as when they had been used previously: that is, to presume the identity of certain symbols and operations. There are, however, necessary conflicts which arise as the relationship is not one of identity. The problem of taking a metaphor literally is augmented by its existence being concealed by the symbolism.

The metaphor involved identifies 2 with the directed number +2 (just as George was identified with the lion), very soon after the introduction of directed numbers. The same notation is used for the subtraction of directed numbers as for the subtraction of (unsigned) counting numbers. The results disagree, however, for whereas $2 - 3$ is impossible within the counting numbers, $+2 - +3 = -1$ within the system of directed numbers. The identification of n with $+n$, which identifies the counting numbers with the positive integers, also identifies the operations.

Discussions of metaphor customarily hinge on the distinction between literal and metaphorical meaning. What are the archetypal situations with the natural numbers and operations on them, which give rise to their meaning? In brief, natural numbers arise as possible answers to the question 'how many?', addition from conflating two distinct sets of objects, and subtraction from physical removal of objects, multiplication from repeated addition and division, an operation related to the actions of both equal sharing and grouping.

What are the equivalent well-springs of meaning for the integers and operations upon them? Under the above identification, the positive whole numbers become the whole numbers by omission. What an omission! One of the most important distinctions is between signed and unsigned quantities and the two distinct types of numbers used to measure them. Too many differences have been elided. Because of the structural metaphor, an extended concept of number is required, one which includes the possibility of directedness. -1 makes little sense as an answer to 'how many?' any more than $+2$ does. With directed numbers, zero need not entail nothing, rather it can be merely an arbitrary reference point from which to measure directed deviations. -1 is commonly read as 'minus one', a verb phrase without a subject. Yet in all previous work, the verb *minus* had always been flanked on both sides by a number. It has now become an object, the noun 'negative one'. (This is the beginning of the deliberate confusion which mathematicians perform which I referred to in Chapter 6 in the context of colour and Moroccan numerals.)

We have switched systems while keeping the language the

same, yet not all of the properties of the old system are carried over by the metaphor +2 = 2 etc., particularly not the specific meanings involved. The central issue is that although the symbolic statements equated by the metaphor may behave in structurally analogous ways, their meanings are seldom compatible. More important, is a straightforward identification of two already extant systems involved, or do the operations on the integers somehow emerge from those on the natural numbers?

Consider the case of multiplication of directed numbers. Is there an appropriate real-world context to which an appeal can be made? One force of the identification of +2 with 2 is to permit the literal meaning of multiplication of counting numbers to be used metaphorically, transferring an old meaning to construct a new one. Thus +2 × +3 takes on the meaning of 2 × +3 or +3 + +3. Having made sense of + in terms of movement to the right – directed – on the number line, the answer +6 is obtained. Multiplication of a positive number by a negative one is feasible without too much stretching, although a negative multiplied by a positive is often approached through commutativity.

Justification of this is fairly flimsy due to the absence of any situational meaning guiding the operations. The reasons become formal, reflecting a desire for logical systems. Not that this is inappropriate *per se*, rather it is so only in the light of the age at which such operations are customarily taught. What can we do with −2 × −3? The metaphor cannot help here and this is, I contend, one of the causes of problems in trying to understand the multiplication of negative numbers.

The algebra of indices produces further instances of precisely this extension of meaning; in one direction to negative numbers and in another to fractions. $a, a^2, a^3, \ldots$ can be viewed as a shorthand notation for repeated multiplication, where the index is the number (in the sense of 'how many?') of times the base number a is to be multiplied together. What sense can be made of a^{-1}? What sort of statement is $a^{-1} = \frac{1}{a}$? The metaphor +2 = 2 etc. suggests we construe a^{+2} as a^2.

Presuming we wish to preserve the law of exponents for counting numbers, namely that $a^m \times a^n = a^{m+n}$, we can give a meaning to $a^0, a^{-1} \ldots$. Thus the metaphoric interpretation has been arranged so that a particular property is transferred. However, just as with falling back on commutativity for the extended concept of multiplication, the question of why this justification should be accepted is a difficult one to answer. In the same way, $a^{\frac{1}{2}} = \sqrt{a}$ can serve as a guide to extending further the range of indices. What about $a^{\sqrt{2}}$ or $a^{\sqrt{-1}}$? 2^2, $2^{\frac{1}{2}}$, $2^{\sqrt{2}}$, 2^π, $2^{\sqrt{-1}}$ all have the same surface structural form, yet each one increases in complexity of meaning, and, frequently, actual computation. The metaphor implicit in the use of the same notation masks such difficulties and changes of meaning.

In mathematics, there are many metaphoric uses of what could be called the fundamental chain of numbers.

$$\text{whole} \subset \text{integers} \subset \text{fractions} \subset \text{real} \subset \text{complex}$$
$$\text{numbers} \qquad\qquad\qquad\qquad\quad \text{numbers} \quad \text{numbers}$$

Despite the many complications of construction, these are viewed as set-theoretic inclusions, although each one requires a structural metaphor. One mathematical strategy is to ascend this chain whenever a use has been found for numbers lower down it.

The focal point of structural metaphors in notation is the assertion of equality. There seem to be at least two major, different uses of the symbol '=' in mathematics. The first is a fairly low-level, naming use, for example, $f(x) = x^2$, where one side has meaning and the other is to be used as a label. Under the influence of computing, where there is an essential distinction between a variable as a name and the value it has, some texts are adopting the notation $f(x) := x^2$ for this type of naming equality. Definitions can be of this sort. Consider the statement (claim?) that $a^{-2} = 1/a^2$. Whenever a^{-2} is encountered, the meaning of the other side is transferred, giving force to Polya's injunction, 'go back to the definitions' (Polya, 1948).

The second use of '=' is more worthy of the name 'theorem', and arises from some kind of identification of meanings, where both sides of the equality have meaning in their own right. For instance, in the case of $\int_a^b f'(x)\, dx = f(b) - f(a)$, the two meanings are being equated under the circumstances detailed in the theorem. Is $2 + 3 = 5$ to be classified in the second sense? What about $2 - 3 = -1$? In the former instance, it is a dynamic equality where the left-hand side has been transformed by some operation and identified as the right-hand side. Many pupils read '=' as 'makes', which has a directional quality to it. $2 + 9$ is not an acceptable answer to what is $4 + 7$ because it still contains an operation, a verb. Leaving a fraction as $\frac{1}{4}$ has an incomplete air about it and the statement that $7 \div 4 = \frac{7}{4}$ is often not recognized by pupils as having achieved anything. In fact, its mathematical status is unclear, in that it is only more than a notational alteration provided $\frac{7}{4}$ is perceived as a mathematical object, and not just an alternative notation for 'seven divided by four'.

Eliminating operation symbols seems to make the answer static, stable and hence, perhaps, permanent. The inability of many pupils to accept as an acceptable object any expression containing an operator sign has deep ramifications in algebra. Booth (1984) and others have documented widespread unwillingness to agree that $2a + 3b$ can be an acceptable answer to a mathematical problem. Algebra relies on the dual perception of all expressions being both objects in their own right and at the same time, comprising a sequence of nested computational instructions.

The symbols $\sqrt{}$ and $\int dx$ also function in a similar way for older students. $\sqrt{2}$ should be evaluated further; it is a call to action, rather than being a nominal expression. The continued presence of operators is indicative of a tension yet to be resolved, which suggests a partial explanation for the difficulty students often encounter on seeing the definition $\log(x) = \int_1^x \frac{1}{t} dt$. This latter expression has the appearance of a naming equality, yet many students feel they already have a meaning for $\log(x)$ and, anyway, why has the integral not been evaluated. These instances provide support for Wittgenstein's perceptive observation that in mathematics processes are always identified with results.

These two sections have both involved the topic of reading written mathematical symbolism, although the latter one was focused primarily on notational metaphor. In the final section, I turn to an apparently new topic, namely meta-linguistics. Yet, in fact, one of the themes which emerged in Chapters 6 and 7 was that of conscious knowledge of various properties of the mathematical writing system itself. This topic will serve as an appropriate resting place in my quest for linguistic aspects of mathematics.

3 Meta-linguistics

In the foregoing chapters, I have explored aspects of mathematical language in its customary forms of speaking and hearing, writing and reading. There is one linguistic component that I have not yet concentrated on – namely knowledge *about* the language itself. The discussions I reported in Chapter 5 on the nature of algebra provided one instance of this where the topic of inquiry was not some mathematical calculation, but the way mathematical symbols themselves were used. In those conversations I was attempting to ascertain the nature and extent of their overt knowledge about algebra and its conventions.

As well as the abilities which permit language to be successfully employed, competent users can monitor themselves and others with regard to well-formed words or even sentences of English. They can express judgments about the acceptability or otherwise of various purported items of that language. In other words, they can reflect on language itself, and provide opinions, for example, on the appropriateness or otherwise of a certain linguistic form or style to a particular social situation. As well as acquiring the ability to use language, children develop definite ideas *about* ordinary language and how it functions, both in its spoken and written forms. While I shall concentrate here mainly on spoken instances, a similar development in meta-linguistic awareness occurs in the case of written language (see Sinclair, 1983).

The term *meta-linguistic* refers to this notion of 'aboutness', contrasting knowledge *of* the language with knowledge *about* language itself. Examples of some of the perceptions which children acquire about natural language include: that word order matters, that the same word can appear with varying endings which make a difference and that certain sentences mean the same as others. However, such meta-linguistic abilities, proposed by Cazden (1972) and others to account for these perceptions and beliefs, are often poorly articulated and not easily accessible to conscious inquiry.

In a book entitled *The Development of Metalinguistic Abilities in Children*, Hakes (1980) claims that, 'the competent speaker-hearer of a language not only produces and understands it but, in addition, has intuitions about it' (p. vi). He argues that meta-linguistic abilities develop separately from ordinary linguistic competences in young children, with the former emerging predominantly during middle childhood; that is, between four and eight years of age. One example of such an ability might be the realization that active and passive sentences are synonymous. The area of meta-linguistic awarenesses is a recent concern of linguists, just as the corresponding concerns of mathematics educators have emerged only in the last fifteen years or so, indicating a broadening in scope of the phenomena of interest of these disciplines. What seems clear is that for such abilities to develop, language (or mathematics) itself must have become an object, whose properties can be examined and reflected upon.

Hakes distinguishes between the comprehension of an utterance and an acceptability judgment of it. He claims the former does not imply the latter and, in fact, asserts that the two are both 'logically and psychologically distinct'. He also notes the initial insensitivity of very young children to word order.

While Hakes is discussing the difficulties and developing awarenesses of very young children, there are major similarities to be explored with anomalies and difficulties which arise much later in the teaching of mathematics. Many older students of mathematics demonstrate an insensitivity with regard to order in or structure of written mathematical expressions. One contributory reason for this may be the fact that commutativity is a common property of mathematical operations, and commutativity apparently denies the importance of 'word'-order. Often the actions of pupils on paper seem to indicate little, if any, awareness of nesting or depth in algebraic expressions, treating for example $\sin(\log(x^2 + 1))$ as a linear string of symbols, to be processed in a left-to-right manner. Another example, cited earlier in the context of an over-generalized rule, can be formulated as a general distributive law for symbols, often cued in by brackets. Thus pupils will often claim that $\sin(A + B) = \sin(A) + \sin(B)$.

In all this work, it is important to be aware that the subjects are being asked about the acceptability of the sentences *qua* sentences, and not about the appropriateness or accuracy of the meaning. Gleitman *et al.* (1972) cite a five-year-old boy who claimed the sentence *I am eating dinner* to be 'silly' (*good* and *silly* being the terms in which the judgments were to be made) and when asked about this, he said that he did not like eating dinner. Similarly, while $7 + 8 = 19$ is grammatically well-formed but unacceptable, $19\ 7 + = 8-$ is ungrammatical as an expression. In order to be able to make such a judgment, divorced from its meaning, the expression has to be perceived as an object in its own right.

The ability to conceive attributes separately and apportion them correctly, both for symbol and object, was explored briefly in Chapter 6. Hakes cites work by Papandropoulou and Sinclair (1974) as evidence of the difficulty young children have with a comparable separation in terms of natural language. They provide evidence of four-to-five-year-old children who, when asked for a long word offered, for instance, *train* (a long object) and when asked for a short word, offered *primula* (a small object). Hakes claims,

> Thus it appears that younger children may be focusing on the 'things' named or described linguistically rather than on the linguistic means used for naming them. Further, it appears that this tendency may disappear (or at least diminish) during middle childhood. . . . The change appears to involve an increasing ability to focus attention on and evaluate the properties of the language *per se*. (p. 28)

Once again, the disappearance of such a tendency in natural language is not the case in mathematics for a number of reasons. In fact, the reverse seems to be true, exacerbated by the common focus in many mathematics lessons on the linguistic means for naming rather than what is named. Due to the abstract nature of mathematical objects, it is less clear what the attributes which are being referred to are, or even what the objects are themselves. Secondly, although conceptual difficulties can and do arise from the practice of treating the symbol as object in mathematics, such a perception is essential in order to become fluent at calculations and manipulations involving symbols.

What ideas do pupils at various ages have about mathematics itself, about how it functions? What are the sources of such knowledge? How does the way they perceive the subject affect their interpretation of what goes on in the classroom? Meta-mathematical knowledge, knowledge *about* mathematics itself, like its natural language counterpart, is often unformulated – diffuse, yet resistant both to change and even conscious access. Exceptions do occur, however, which afford us glimpses into the perceptual worlds of others. Johnson (1983) attempts to find a representation which reflects eleven-year-old Karen's 'knowledge of sums', which Johnson has elicited from her over the course of a number of conversations about mathematics.

Hughes (1986), working with young (three-to-seven-year-olds) children's idiosyncratic attempts at representations of numbers

and operations and their beliefs about the marks they were asked
to make, raises a plethora of issues to do with the purposes of
written language which adults tend to take for granted. For
instance, is recognizability or retrievability important? Is there
the possibility that others might be able to use them in similar
ways? Classifying the resulting representations into the categories
of *idiosyncratic*, *iconic*, *pictographic* and *symbolic*, Hughes
comments that, 'On the whole, the children were able to identify
symbolic, pictographic and iconic responses, but found their own
idiosyncratic responses as inscrutable as an adult world.' In
conclusion, he stresses the importance of ascertaining pupil
beliefs about various processes, in this particular instance those
about written symbolism and the uses to which it might be put,
before teaching the conventional notation.

A strong example of a pupil's ideas about mathematics is
provided by Erlwanger (1973). Erlwanger's work is firmly rooted
in a tradition which holds that children's errors are often both
comprehensible and informative. (Another exponent of this view
is Ginsburg (1977, p. 128), who observes that 'typically children's
errors are based on systematic rules.') In order to make sense of
the regularities present in many children's errors, it is important
to try to gain access to the processes with which they are
operating on the symbols. Such reflection, while almost impossible at the age when children are acquiring the bulk of their
natural language, is more feasible in the mathematical context.
One of the themes of this section has been that some of the same
difficulties that are evident in children learning mathematics have
already been overcome by those same pupils in the context of
their first language acquisition.

Erlwanger's concern is 'to explore each child's ideas, beliefs
and views about mathematics and the process of learning
mathematics'. He held a number of conversations with Benny, an
eleven-year-old pupil who was scoring highly within the individualized instruction system operating in his class. Although his
initial mission at the school was to be of assistance to pupils
requiring remedial instruction, Erlwanger never corrected any of
Benny's errors. While this neutrality of response could have been
interpreted by Benny as Erlwanger's tacit acquiescence in the
validity of his responses, it became clear that even when
Erlwanger attempted to engage himself in remedial work, the
stability of Benny's responses showed them to be remarkably

resistant to change. Erlwanger noted, 'He does not alter his answers or his methods under pressure', and commented earlier that, 'In interviewing him at this stage, I did not attempt to teach him or even to hint as to which answers were correct. He did not ask for that either.' One reason for this might have been that Benny was unused to talking to anyone about mathematics, his sole source of authority being the answer key. One of the strongest complaints which Erlwanger levels at this form of individualized activity is that it renders mathematics an anti-social activity.

Erlwanger talked to Benny about the work the boy was currently engaged in (fractions and decimals), rather than introduce his own problems. Benny, it transpired, had surprisingly well-articulated perceptions about how mathematics operates.

E: How would you write $\frac{2}{10}$ as a decimal fraction?
B: One point two (writes 1.2)
E: And $\frac{5}{10}$?
B: 1.5
. . .
E: And $\frac{4}{11}$?
B: 1.5
E: Now does it matter if we change this and say that it is eleven fourths? (writes $\frac{11}{4}$)
B: It won't change at all; it will be the same thing . . . 1.5

Later discussion elicits:

B: In fractions we have a hundred different kinds of rules . . .
E: Would you be able to say the hundred rules?
B: Ya . . . maybe, but not all of them.

Benny has many ways of coping with errors and with the fact that there are many different ways of writing fractions or decimals.

B: Then I get it wrong because they [aide and teacher] expect me to put ½. Or that's one way. . . . But if I did that also I get it wrong. But all of them are right.
E: Why don't you tell them?
B: Because they have to go by the key . . . what the key says.

I don't care what the key says; it's what you look on it. That's why kids nowadays have to take post-tests. That's why nowadays we kids get fractions wrong.
    . . .
B: Wait I'll show you something. This is a key. If I ever had this one (2 + .8) . . . actually if I put $2\frac{8}{10}$, I get it wrong. Now down here, if I had this example (2 + $\frac{8}{10}$) and I put 1.0, I get it wrong. But really they're the same, no matter what the key says.

Benny mainly views fractions and decimals as symbols and operations on them merely as particular rules of combination of the symbols and he is quite articulate about them. Rules are important 'because if all we did was to put any answer down, [we would get] a hundred every time. We must have rules to get the answers right.' The symbols are the objects with which mathematics deals and the many equivalences he invokes rule out contradictions entering from work with physical apparatus or diagramatic representations. Erlwanger's intention in his article is to turn our attention toward powerful guiding concerns such as children's conceptions of mathematics; in other words, what sort of activity they believe themselves to be engaged in.

These attitudes and conceptions of knowledge, due to their personal, idiosyncratic nature, are scarcely amenable to agrarian statistical models.

One incident with one child, seen in all its richness, frequently has more to convey to us than a thousand replications of an experiment conducted with hundreds of children. Our pre-occupation with replicability and generalizability frequently dulls our senses to what we may see in the unique unanticipated event that has never occurred before and may never happen again (Brown, 1981, p. 11).

The case of Benny shows clearly that children do have conceptions of mathematics, and their beliefs can colour their views and beliefs about what happens in mathematics classes.

Summary

Although many of the symbols employed in mathematics come from various alphabets, there is no sense in which mathematical symbolism can be said to be alphabetic. Consequently, a phonic method is quite inappropriate to the problem of reading mathematical symbolism. Mathematical notation is trans-national, in that the same piece of text can be presented to mathematically literate individuals in a wide range of countries, who will then read it in(to) their own languages, which can have a wide range of different grammatical structures (e.g. Finnish, Rumanian, Japanese, Russian, . . .). If mathematical symbolism can be claimed to be akin to a written language, there is no spoken counterpart and there are no native speakers.

Many different notions are designated by the same symbol. Some of these are the mathematical writing systems version of *homographs*, that is, words which mean different things but are not distinguished orthographically from one another. Thus we saw in Chapter 6 that the colon is put to a number of different uses in mathematics. However, there are some symbols where the meanings are closely related but not identical, and for some of these the notion of structural metaphor (as outlined in Chapter 4) is appropriate. As I indicated, there is a problem of pupils not realizing that different (if related) concepts are being employed. There is no trace in the symbolism to indicate that a metaphoric usage is being employed. A metaphor skims over a lot and a polished notation permits this riding on the surface. The same words and symbols are used throughout.

I ended this chapter by looking at the concept of meta-linguistic knowledge. At a fairly young age, children develop sufficient knowledge about language to distinguish, for example, properties of the symbols of language from those of the meanings to which they are linked. Why is this so much harder in mathematics? One recurring theme throughout this book has been the symbol/referent distinction and how mathematicians successfully blur it for their own ends. They merge the two, but can separate them at will. As with my earlier remarks about extra-mathematical metaphors, it is important for our pupils to be able to perform the same separation. 'The symbol is the object' is an extremely powerful metaphor at the heart of mathematics, but with this power comes the potential for the destruction of meaning.

9

Mathematics as a language?

> To understand a language means to be master of a technique.
>
> Ludwig Wittgenstein, *Philosophical Investigations*

My starting point for this book was the exploration of the often heard but seldom seriously explored claim that mathematics is a language. It is one of a number of phrases (another being *the syntax of mathematics*) used informally to try to come to grips with language issues in mathematics. The question of possible interrelations between mathematics and language is similar in form to the theme discussed by Virginia Woolf in *A Room of One's Own*. In that book, she explored various possible interpretations of the phrase *women and fiction*, a topic she had been invited to address.

> The title *women and fiction* might mean, and you may have meant it to mean, women and what they are like; or it might mean women and the fiction they write; or it might mean women and the fiction that is written about them; or it might mean that somehow all three are inextricably mixed together and you want me to consider them in that light. (Woolf, 1945, p. 5).

While the parallel is not exact, there is a similar range of interpretations possible for the theme of mathematics and language. The first is that language and mathematics are seen as co-existing entities, to be placed side by side, then compared and contrasted. The links, if any, are ones of similarity. However, thinking about language and mathematics as separate entities seems inappropriate. It is how language is modified as a result of attempting to communicate mathematical ideas and perceptions which is of far greater import. That there are any links of value is by no means a universal belief. Indeed, as eminent a mathematician as the intuitionist Felix Brouwer insisted that, 'mathematics is a languageless activity of the human mind.'

A second style of interpretation gives rise to the notion of *the mathematics of natural language*. This idea leads in one direction to the formal theory of transformational grammar originating with Noam Chomsky, and in another to the work of Caleb Gattegno. Chomsky's (1957, 1965) theory attempts to provide a means of analysing sentence structure. In particular, it indicates how any grammatical expression may be derived using a system of abstract rules to generate templates of sentences into which the particular words may then be inserted. 'Language is mathematical in nature' might be an appropriate description of this viewpoint, and Chapter 7 contained a discussion of the relation of Chomsky's notion of linguistic transformation to that of transformations in mathematics.

Gattegno, while not dismissing the work of such linguists, has an alternative interpretation in mind for the expression *the mathematics of natural language* – one which he presents as a challenge for mathematicians themselves:

> When linguists use mathematics to describe their awareness, they borrow models from the existing literature, models which were originally produced for an entirely different purpose and which, more often than not, fail to provide a deeper understanding of the situation. This use of existing mathematics seems to linguists as legitimate as it does to engineers, physicists or biologists. What I am proposing here is the reverse process: the creation of a new mathematics and its problematics for the purpose of formalising the awareness of the dynamics of speech.
>
> One of the difficulties resides in the fact that the grasp of meanings precedes verbalization and that words *per se* are not the message, but only one of the possible vehicles for the message. Hence the mathematics of speech cannot in fact result from the analysis of sentences, but can only arise through an understanding of the mysterious ways in which these carry meanings (Gattegno, 1970b, p. 137).

Gattegno has worked intensively on questions concerning the learning of mathematics and foreign languages from a viewpoint dominated by an immense respect for the intellectual powers of children evidenced in their mastery of their first language. He perceives these powers as mathematical in nature and has attempted to develop teaching materials and methods which draw upon them. Although the challenge he has offered was not directly addressed or taken up here, it is nonetheless one worthy of attention.

A converse possible interpretation of this second suggested relationship between mathematics and language is *the language of mathematics*, that is the language in which mathematics is spoken or written. This phrase is interpretable in a number of ways in the light of this book.

A third distinct possibility of interpretation can best be indicated by means of the expression *mathematics as a language*. In order to make this clearer, let me summarize the possibilities discussed so far. There seem to be three common levels of relationship which are signalled in academic book titles. They can be exemplified in the current context as follows:

(i)   mathematics and language (X and Y)

(ii) the mathematics of language (the X of Y) or
      the language of mathematics (the Y of X)

(iii) mathematics as a language (X as Y)

Actual illustrations of these title structures include the books *Chance and Necessity* (Monod, 1972), *The Politics of Pure Science* (Greenberg, 1967) and the more recent *Language as Social Semiotic* (Halliday, 1978).

In the first possibility, X and Y, the two elements are free-standing and are to be related, compared and contrasted by juxtaposition. In the second, the X of Y, one element (X) is heavily subordinate to the other (Y), as can be illustrated by means of reversing the position of the two terms. The title of a survey paper, *The Architecture of Mathematics*, indicates how the second title structure referred to earlier (the Y of X) can be inherently metaphorical. (Mathematics is a large 'building' and hence can be said to have an architecture.) While there is a definite directionality in the third structure (X as Y), the two components are to be fused. It is this third possibility, that of construing mathematics in linguistic terms, which has formed one central theme of this book.

1 How to pursue a metaphor systematically

Seeing one thing in terms of another is the province of metaphor. My underlying theme was one of metaphor, that subtle, ill-understood process which permits meanings to be constructed and compared by means of creative mental and verbal juxtapositions. Metaphor is not solely, or even primarily, a decorative device of poetry. Rather it is one of the central linguistic strategies at our disposal with which to create sense of the world. Seeing through metaphors is essential in both senses; we cannot do without this process, yet it is important to be aware when we are using them.

(I have been accused of seeing metaphors everywhere. There is a delightful cartoon (Koren, 1981) comprising a bearded, bespectacled male in a country scene where all the objects carry labels. The tree wears a sign 'metaphor for growth and change', the fence by the side of the road 'metaphor for limits', while the two adjacent flowers are to be seen as a 'metaphor for love'. It provides a visual instance of seeing both the general and the abstract in the particular.)

I want to discuss briefly the way in which I have worked with the apparent proposition 'mathematics is a language', as I believe it to have wider ramifications for mathematics education as a whole. I have treated it as an instance of a systematic, *structural* metaphor of the type discussed at length by Lakoff and Johnson (1980). (Note that their use of the term *structural* bears no relation to the mathematical use I coined in Chapter 4.)

Lakoff and Johnson cite the expression *argument is war* as an example of this particularly central class of metaphors. Conversing in everyday life about arguments we say things like:

You outmanoeuvred me
I won that argument
He attacked my every weak point
I lost the initial encounter with him
That position is indefensible
If you use that strategy, you're asking for trouble
Let me try a new line of attack
. . .

There are a number of points to note about this example. Firstly, these are everyday expressions. Secondly, the above metaphor (succinctly condensed into the phrase 'argument is war') is *systematic*, rather than a one-off expression, in the sense that there is a systematic transfer of expressions from those used to describe warfare to the field of verbal arguments.

Not all metaphors are systematic. In the case of the expression 'the leg of the table', the *eye* of the table or the *arm* of the table are not familiar expressions (although we do talk of the *head* or *foot* of the table), suggesting that 'a table is a body' is not a systematic metaphor in our culture. Even in the case of the head and the foot of the table, although the head and the foot are at opposite ends, the head is as close to the legs as the foot is. This indicates a relationship between the parts of a body which does not carry over to the new context of use. Some of the important ramifications for mathematics of this phenomenon of certain relations failing to be conveyed by a metaphor (even a systematic one) were explored in Chapter 4.

Thirdly, metaphor should not be thought of as a purely verbal phenomenon. One instance of the effect of the metaphoric expression *the leg of the table* might have been the Victorian habit of covering armchair and piano legs with ruffs, as it was considered indelicate for legs to be uncovered. Thus, certain appropriate actions were transferred as a result of the perception engendered by this isolated metaphoric usage.

As a clearer example of a metaphoric *action*, a colleague was walking down the corridor of an English department and saw a friend talking to a third person. The friend was swinging a coffee cup gently from side to side. The visual impression evoked the image of a beggar, and so on passing, to their great surprise, my colleague dropped a 10p coin in the empty cup. No word had been spoken. Rather a metaphoric thought had been embodied in a sequence of *actions* rather than words.

These latter examples provide instances of metaphor functioning at the level of thought and being manifested in actions based upon them. Lakoff and Johnson assert that, 'Metaphor is pervasive in everyday life, not just in language but in thought and action. Our ordinary conceptual system, in terms of which we both think and act, is fundamentally metaphorical in nature.' The *perception* therefore, rather than the linguistic form, is of central importance, and one important disputed question is the extent of the reverse influence. What is the connection between how things are described and how they are seen? That is to say, does the fact that different cultures express things verbally in particular ways affect how these things are seen, and hence, for instance, what options are perceived as open? (A national newspaper published an article touching on this theme, exploring it in the context of the economic metaphors of illness and health employed by Prime Minister Margaret Thatcher (Beattie and Speakman, 1983).)

Finally, in summary, Lakoff and Johnson claim (p. 6) that, 'The essence of metaphor is understanding and experiencing one thing in terms of another.' 'Argument is war' is systematic and therefore, they argue, the existence of this pervasive metaphor considerably determines how we view arguments. The concept of *argument* in our culture is partially structured in terms of that of *war*. Yet despite the widespread use of such borrowed expressions, Lakoff and Johnson point out, argument is not the same thing as war; verbal discourse is different from armed conflict.

It is the juxtaposition of two different elements which is crucial. If one is actually confused with the other, there is no work the metaphor can do. Metaphors can conceal their existence by masquerading as literal speech. But once the metaphor has been recognized explicitly, we are suddenly liberated. For instance, it is now possible to conceive of argument in different terms. Lakoff and Johnson invite us to break apart the metaphor of seeing argument as war by means of imagining a culture in which the dominant image is of argument as dance.

What I hope I have illustrated in this book is how by successfully using one concept to structure another metaphorically, a systematic domain for inquiry can be created. To do so for mathematics and language enables certain questions to be formulated which may not otherwise have been noticed as being important, reasonable or even coherent.

2 The teaching of mathematics

I have pursued in this work the metaphor of seeing mathematics as a language. This identification carries with it an implied view of mathematics teaching as language teaching which may colour what is seen when looking at a mathematics classroom and at mathematics itself through linguistic eyes. I indicated at the outset that Gattegno and others would like to see the learning of mathematics in terms of the same processes as first language acquisition. However, it might also be useful to see mathematics teaching in terms of learning a foreign language. In particular, if it is useful to see the learning of mathematics as akin to learning a foreign language, what implications for the teaching of mathematics can be drawn from the many innovations in foreign language teaching methods which have occurred recently. (See e.g. Brumfit and Johnson, 1979.)

Until fairly recently, foreign language teaching in many secondary schools tended to be a formal and deliberate affair, which was largely rule-based: that is to say, rules were given as to how to form the passive, the negative, the future tense, etc. These rules were then practised in production. The utilitarian view of being able to do something with language was largely missing, and language teaching and learning was consciously self-conscious. That is, the language itself was the centre of attention, the medium rather than the message.

The last fifteen years, however, have seen a rapid upsurge in interest in the notion of communicative language teaching, that is to say, attempting to teach communicative competence in a language. Communication involves exchange of meanings. Recall from Chapter 1 that the general notion of communicative competence involves knowing how to use language to communicate in various social situations. In other words, it requires an awareness of the particular conversational or written, context-dependent conventions operating, how they influence what is being communicated and how to employ them appropriately according to context. The preceding argument raises the question of whether this is the case with mathematics teaching, particularly when teachers ask pupils to record. If the notion of communicative mathematics teaching were to gain ground, perhaps many more pupils would find mathematics an engaging and rewarding study.

Mathematics teaching still suffers from the same difficulties as the former teaching strategy for foreign languages. The attention of both teacher and learner is most frequently on the *form* of the mathematical statements themselves, rather than on the ideas expressed by them. The Cockcroft report (para. 306) claims that, 'language plays an essential part in the formulation and expression of mathematical ideas'. Unfortunately, it is seldom the pupil's expression of the pupil's formulation of the pupil's ideas. A similar idea is expressed in the phrase 'doing someone else's algebra rather than your own' (Mason *et al.*, 1985). Just as with

communicative language teaching, for communicative mathematics teaching, the pupils doing the communicating must have
something they wish to express.

Instead, notation and higher-level *structural* concerns of the
subject (the 'spelling' and 'grammar' perhaps) are still frequently
uppermost as the focus of attention, with a concomitant emphasis
on the 'right' way of doing things. One focal concern of all
involved with mathematics education should be to ascertain how
to deny symbols pride of place as the objects of mathematical
inquiry, a place they have the tendency to usurp by default. This,
in part, is due to an inability or unwillingness to talk about what
might take their place. Many pupils sitting looking blankly at a
page of symbols will all too readily attest to the reality before
their eyes. By conversing about the objects and situations
mathematical questions are inquiring about, mathematical
symbols can be returned to the role of 'carrier', of medium rather
than message.

Instead, rules are given to be practised – exposition/exercise in
the terminology of the Cockcroft report (para. 243) – and the
outcome is judged right or wrong. Mathematics is commonly
viewed as a subject with the clearest guide to right and wrong
responses and such judgments are regularly exercised as part of
the teacher's function. Yet if learning mathematics were seen in
terms of acquisition, rather than overt learning, alternative
strategies with regard to error correction might ensue. One
instance brought about by seeing mathematics teaching as
language teaching could therefore be a change in focus from the
study of a rule-governed abstract system, with an emphasis on
written forms, to one of the acquisition of communicative
competence about certain objects, situations and phenomena,
with a concomitant oral emphasis.

As with young children learning to read and write, gaining
mastery over mathematical symbols and allowable transformations of them can fully occupy pupils. There is a period in the life
of children when they are unable to control spoken words – one
they pass through. There is a later period when they are unable
to control written language – almost all children overcome this
barrier, to a greater or lesser extent. There is a still later period
when pupils are unable to control mathematical language,
whether spoken or written – very few succeed in this challenge to
reach any sort of mastery and fluency.

What is the aim? Subsequently, it must be hoped, the symbols can sink into the background of attention, bearing the same relation to the mathematical meanings they convey as words do in ordinary language. The situation that some foreign language learners attest to, namely one where they are rendered almost incapable of speaking due to grammatical concerns being uppermost, occurs daily in many mathematics classes. The above expressed hope may, however, be a forlorn one. This is because the symbol-transforming aspect of mathematics, which has proven to be such a powerful analytical tool, relies for fluency on the ability of the transformer to disconnect the symbol from its referent, and work with the symbol alone.

The metaphor of 'the symbol is the object' (which mathematics has made its own), although powerful, is two-edged. A colleague, when working on a problem commented that, 'I'm now just fiddling with the symbols and doing nothing that has any relation to the problem'. General permissions involving symbolic trans-formations which are expressed in the so-called 'rules' of algebra allow solutions to be obtained where the corresponding trans-formations of the referent objects are opaque in the extreme. Yet encouraging people to move into this blind symbol-manipulating mode can cut them off from the semantic content of mathematics from which it derives its purpose and meaning. This tension provides yet another dilemma for the mathematics teacher.

But what of the future? There is at least one major linguistic issue concerning mathematics which I have not addressed in this book, namely the complex social and intellectual problems arising from attempting to learn mathematics through a language which is not one's native language (see UNESCO, 1975 and Dawe, 1983). This situation exists for an enormous number of the world's pupils and is becoming of increasing relevance and concern in Britain, particularly in urban areas which contain sizeable communities for whom English is a second language. The issue raises important questions about the nature of the dependence of mathematical ideas on fluency in the language of instruction, the question of translation between languages of widely differing structure and the extremely vexed question of relations between language and thought itself (see Vygotsky, 1962 and Higginson, 1978).

3 The goals of mathematics education

Each chapter of this book contains unanswered questions. In the words of Wallace Shawn in the film *My Dinner with André*, 'Asking questions relaxes me.' I would defend the discriminate posing of questions as a valuable activity in its own right, despite the following polemical attack on the general enterprise.

As is perhaps obvious, Morris Zapp had no great esteem for his fellow-labourers . . . They liked to begin a paper with some formula like, 'I want to raise some questions about so-and-so', and seemed to think they had done their intellectual duty by merely raising them. This manoeuvre drove Morris Zapp insane. Any damned fool, he maintained, could think of questions; it was the answers that separated the men from the boys. If you couldn't answer your own questions it was either because you hadn't worked on them hard enough or because they weren't real questions. In either case you should keep your mouth shut. One couldn't move in English studies these days without falling over unanswered questions which some damn fool had carelessly left lying about. (Lodge, 1978, p. 45)

In a book entitled *The Art of Problem Posing*, Brown and Walter (1983) offer 'a celebration of creative questioning' and a number of general techniques by means of which questions in mathematics may be posed. The judgment of the value of the questions is a later activity best separated from the generation of the questions themselves. But they draw emphatic attention to the activity of posing questions as a worthy activity in its own right.

There has been considerable informal debate within the mathematics education community as to whether or not mathematics education can reasonably be claimed to constitute a discipline in its own right. If so, wherein lies its coherence? Is it derived from the particular phenomena of interest themselves, or the methods of approach to inquiring into them? One alternative way of conceptualizing a discipline is in terms of the questions asked, rather than the ways with which they are dealt.

The excitement and the value of this present volume is that it shows how productive it can be to think in terms of questions rather than territories; to think in terms of what kinds of evidence are relevant to the enquiry rather than what kinds of data can best be used as numerical props to the argument; and to think in terms of where the logic of enquiry must lead rather than what areas are considered safe to enter. (P. Doughty, Introduction to M. A. K. Halliday, (1975b) *Learning How to Mean*)

David Wheeler has attempted an informal garnering of questions of Hilbertian status from certain members of the mathematics education community. Some of the individual contributions to his survey can be found in an issue of the journal *for the learning of mathematics* (*4*(1), 1984), which Wheeler edits. A general overall summary of the responses is contained in Wheeler (1983b).

The philosophers have appropriated the term *meta-mathematics* and used it to describe a very narrow mathematical activity, namely the study, including the proving of theorems, about logical systems such as mathematical theories themselves. It comprises a formal study of mathematical language. It was in this sense, I believe, that Wheeler condemned the region *Mathematics is a Language* in terms of it being uninhabited. Part of what I am seeking to do is to reclaim the term *meta-mathematics* for mathematics education, and also partially to define the scope of the latter enterprise in terms of the former. I have tried to show in the course of this book how, by seeing mathematics in terms of language, this perception can shed some light on issues of teaching and learning mathematics. In the course of so doing I have offered a non-standard instance of Thom's already-cited claim, that 'all mathematical pedagogy, even if scarcely coherent, rests on a philosophy of mathematics.' For me, one goal of mathematics education should be to ascertain and explore ways of seeing mathematics which provide insight into its learning and teaching. In that sense, mathematics education can be seen as meta-mathematics.

As a final instance of such a perception generating questions, there is the notion of 'the literature', both in mathematics and English. The last couple of years have seen a fiery debate over what should constitute the study of English at university, though there seems to be little dissension among English scholars from the view that literary criticism *of some form* should be the central focus of study. *Whose* literary criticism is more the question at stake. Where is the corresponding 'critical theory' debate in mathematics? (See Pimm, 1982.) Why is the study of mathematics not more like the study of English? What would a theory of mathematical literary criticism look like? (For one rare instance, see Lakatos, 1976.) Once again, the metaphor 'mathematics is a language' enables a question to be asked, one I think of considerable import for tertiary mathematics education, which may not otherwise be perceived as relevant or even reasonable.

In conclusion

In what sense, then, can it be claimed that mathematics is a language? Mathematics is not a natural language in the sense that English and Japanese are. It is not a 'dialect' of English (or any other language) either. Many natural languages have developed registers which allow discussion about mathematical concerns to take place, and the fact that it is *mathematics* that is under discussion has also shaped the language used. Learning to speak, and more subtly, learning to *mean* like a mathematician, involves acquiring the forms and the meanings and ways of seeing enshrined in the mathematics register.

Mathematics has a writing system which is complex and rule-governed and the metaphoric expression *the syntax of mathematics* proved to have considerable force in describing the rule-governed manipulations of symbols which form such a part of the mathematician's art. The symbolic aspect of written mathematics, together with the mathematician's encouragement of confusion between symbol and object, in addition to the abstract nature of mathematical objects themselves, all add to the perception of mathematics being a language. As I hope I have shown, this way of seeing can have its uses, but also has its dangers. By exploring in detail the extent of some of these perceptions, I have tried to put some of the problems of learning and teaching mathematics into a new light.

Bibliography

Books

Acland, R. (ed.) (1984) *Investigating Talk in Cumbrian Classrooms*, Longman, York.

Adda, J. (1982) 'Difficulties with mathematical symbolism: synonymy and homonymy', *Visible Language*, *16* (3), pp. 205–14.

Akmajian, A. and Heny, F. (1975) *An Introduction to the Principles of Transformational Syntax*, MIT Press, Cambridge, Mass.

Ardrey, R. (1967) *The Territorial Imperative*, Collins, London.

Austin, J. L. and Howson, A. G. (1979) 'Language and mathematical education', *Educational Studies in Mathematics*, *10*(3), pp. 161–97.

Barnes, D. (1969) 'Language in the secondary classroom', in Barnes, D. *et al.*, *Language, the Learner and the School*, Penguin, Harmondsworth.

Barnes, D. (1976) *From Communication to Curriculum*, Penguin, Harmondsworth.

Beattie, G. and Speakman, L. (1983) 'A load of old trope', *Guardian*, 4 March.

Beeney, R. *et al.* (1982) *Geometric Images*, Association of Teachers of Mathematics, Derby.

Bell, M. (1982) The News, BBC1-TV, 16 February.

Booth, L. (1984) *Algebra: Children's Strategies and Errors*, NFER-Nelson, Windsor.

Boyer, C. B. (1968) *A History of Mathematics*, Wiley, New York.

Brand, T. E. *et al.* (1969) *Contemporary School Mathematics 4*, Edward Arnold, London.

Brice, W. C. (1976) 'The principles of non-phonetic writing', in W. Haas (ed.) *Writing Without Letters*, Manchester University Press, Manchester, pp. 29–44.

Brookes, W. M. (1970) Preface to *Mathematical Reflections*, ATM members (eds), Cambridge University Press, Cambridge.

Brown, G. (1982) 'The spoken language', in Carter, R. (ed.), *Linguistics and the Teacher*, Routledge & Kegan Paul, London.

Brown, S. I. (1978) *Some Prime Comparisons*, National Council for Teachers of Mathematics, Reston, Virginia.

208

Brown, S. I. (1981) Sharon's Kye, *Mathematics Teaching*, *94*, March, pp. 11–17.

Brown, S. I. and Walter, M. I. (1983) *The Art of Problem Posing*, Franklin Institute Press, Philadelphia, Penn.

Brumfit, C. J. and Johnson, K. (1979) *The Communicative Approach to Language Teaching*, Oxford University Press, Oxford.

Bundy, A. (1984) *Computer Modelling of Mathematical Reasoning*, Academic Press, London.

Carroll, L. (1865) *Alice's Adventures in Wonderland*, Macmillan, London.

Cazden, C. (1972) *Child Language and Education*, Holt, Rinehart & Winston, New York.

Chomsky, N. (1957) *Syntactic Structures*, Mouton, Dordrecht.

Chomsky, N. (1965) *Aspects of the Theory of Syntax*, MIT Press, Cambridge, Mass.

Clark, H. H. and Clark, E. V. (1977) *Psychology and Language*, Harcourt Brace Jovanovich, New York.

Coghill, V. (1978) 'Infant School Reasoning', mimeo, cited in Corran, G. and Walkerdine, V. *The Practice of Reason*, University of London Institute of Education, London.

Cornu, B. (1981) 'Apprentissage de la notion de limite: modèles spontanés et modèles propres', *Proceedings of the fifth PME Conference*, Grenoble, France, pp. 322–326.

Daife, A. (1983) *Entiers Relatifs: Modèles et Erreurs*, unpublished M.Sc. thesis, Université de Quebec à Montreal, Canada.

Davis, R. B. (1966) *Explorations in Mathematics*, Addison-Wesley, Menlo Park, Ca.

Dawe, L. (1983) 'Bilingualism and mathematical reasoning in English as a second language', *Educational Studies in Mathematics*, *14*(4), pp. 325–53.

Dodgson, S. (1898) *The Life and Letters of Lewis Carroll*, Collingwood, London.

Easley, J. and Easley, E. (1982) 'Mathematics can be natural', CCC Research Report #23, Bureau of Educational Research, University of Illinois, Urbana-Champaign.

Eliot, T. S. (1974) *Little Gidding*, in *Collected Poems 1909–1962*, Faber & Faber, London.

Erlwanger, S. (1973) 'Benny's conception of rules and answers in IPI mathematics', *Journal of Children's Mathematical Behaviour*, *1*(2), pp. 7–26.

Farnham, D. (1975) 'Language and mathematical understanding', *Recognitions*, *5*.

Feynman, R. P. (1965) 'New textbooks for the "new" mathematics', *Engineering Mathematics*, *28*(6), pp. 9–15.

Flanders, N. (1970) *Analysing Teaching Behaviour*, Addison-Wesley, London.

Fowler, D. H. (1973) *Introducing Real Analysis*, Transworld, London.

Fraleigh, J. B. (1967) *A First Course in Abstract Algebra*, Addison-Wesley, Reading, Massachusetts.

Gattegno, C. (1969) *Reading with Words in Colour*, Educational Explorers, Reading.

Gattegno, C. (1970a) *What We Owe Children*, Outerbridge & Dienstfrey, New York.

Gattegno, C. (1970b) 'The human element in mathematics', in ATM members (eds), *Mathematical Reflections*, Cambridge University Press, Cambridge, pp. 131–7.

Ginsburg, H. (1977) *Children's Arithmetic: the Learning Process*, Van Nostrand Reinhold, New York.

Gleitman, L. R. *et al.* (1972) 'The emergence of the child as a grammarian', *Cognition*, *1*, pp. 137–64.

Greenberg, D. S. (1967) *The Politics of Pure Science*, New American Library, New York.

Grenfell, J. (1977) *George – Don't Do That*, Macmillan, London.

Hakes, D. T. (1980) *The Development of Metalinguistic Abilities in Children*, Springer-Verlag, Berlin.

Halliday, M. A. K. (1975a) 'Some aspects of sociolinguistics', in *Interactions between linguistics and mathematical education*, UNESCO, Copenhagen, pp. 64–73.

Halliday, M. A. K. (1975b) *Learning How to Mean*, Edward Arnold, London.

Halliday, M. A. K. (1978) *Language as Social Semiotic*, Edward Arnold, London.

Hart, K. M. (1981) *Children's Understanding of Mathematics: 11–16*, John Murray, London.

Harvey, R. (1983) ' "I can keep going up if I want to": one way of looking at learning in mathematics', in Harvey, R. *et al.*, *Mathematics* (Language, Teaching and Learning #6), Ward Lock Educational, London.

Hesse, M. (1966) *Models and Analogies in Science*, University of Notre Dame Press, Illinois.

Hewitt, D. (1985) 'Equations', *Mathematics Teaching*, *111*, June, pp. 15–6.

Higginson, W. (1978) 'Language, logic and mathematics: reflections on aspects of educational research', *Revue de Phonetique Appliquée*, *46–47*, pp. 101–32.

Higginson, W. (1980) ' "Berry undecided": a digital dialogue', *Mathematics Teaching*, *91*, June, pp. 8–13.

Holt, J. (1970) *How Children Learn*, Pelican, Harmondsworth.

HMSO (1982) *Mathematics Counts*, HMSO, London.

HMSO (1985) *Mathematics from 5 to 16* (Curriculum Matters 3), HMSO, London.

Hudson, R. (1984) *The higher-level differences between speech and writing*, CLIE working paper #3, baal/lagb, London.

Hughes, M. (1986) *Children and Number*, Basil Blackwell, Oxford.

James, N. and Mason, J. (1982) 'Towards recording', *Visible Language*, 16(3), pp. 249–58.

Jaworski, B. (1985) 'A poster lesson', *Mathematics Teaching*, 113, December, pp. 4–5.

Johnson, N. (1983) 'Karen's knowledge of sums', in Bliss, J. *et al.* (eds), *Qualitative Data Analysis for Educational Research*, Croom Helm, London.

Kane, R. B. *et al.* (1974) *Helping Children Read Mathematics*, American Book Company, New York.

Kerslake, D. (1979a) 'Talking about mathematics', in Harvey, R. *et al.*, *Mathematics* (Language, Teaching and Learning #6), Ward Lock Educational, London.

Kerslake, D. (1979b) 'Visual mathematics', *Mathematics in School*, 8(2), March, pp. 34–5.

Klemme, S. L. (1981) 'References of speech acts as characteristics of mathematical classroom conversation', *Educational Studies in Mathematics*, 12(1), pp. 43–58.

Kline, M. (1973) *Why Johnny Can't Add*, St Martin's Press, New York.

Koren, E. (1981) *Well, there's your problem*, Penguin, Harmondsworth.

Lakatos, I. (1976) *Proofs and Refutations*, Cambridge University Press, Cambridge.

Lakoff, G. and Johnson, M. (1980) *Metaphors We Live By*, University of Chicago Press, Chicago.

Lehrer, A. (1974) *Semantic Fields and Lexical Structure*, North-Holland, Amsterdam.

Lemay, F. (1983) 'Some lessons in the sun', *Mathematics Teaching*, 105, December, pp. 28–31.

Lively, P. (1974) *The House in Norham Gardens*, Heinemann, London.

Lodge, D. (1978) *Changing Places*, Penguin, Harmondsworth.

Lyons, J. (1984, 2nd edn) *Chomsky*, Fontana, London.

Mason, J. (1980) 'When is a symbol symbolic?', *for the learning of mathematics*, 1(2), pp. 8–12.

Mason, J. and Pimm, D. (1984) 'Generic examples: seeing the general in the particular', *Educational Studies in Mathematics*, 15(3), pp. 277–89.

Mason, J. *et al.* (1985) *Routes to Algebra* (PM641), Open University Press, Milton Keynes.

Mason, J. and Pimm, D. (1987) *Discussion in the Mathematics Classroom* (PM644), Open University Press, Milton Keynes.

Maxwell, J. (1985) 'Hidden messages', *Mathematics Teaching*, 111, June, pp. 18–19.

Milligan, S. (1972) *The Goon Show Scripts*, The Woburn Press, London.

Mmari, G. R. V. (1975) 'Tanzania's experience in, and efforts to resolve, the problem of teaching mathematics through a foreign language', *Language and the Teaching of Science and Mathematics, with special reference to Africa*, CASME, UNESCO, Paris.

Monod, J. (1972) *Chance and Necessity*, Collins, London.

Nesher, P. (1984) Talk given at BSPLM meeting, Chelsea College, London.

Neugebauer, O. (1969) *The Exact Sciences in Antiquity*, Dover, New York.

Nolder, R. (1984) *Metaphor: Its Influences on the Teaching and Learning of Mathematics in the Secondary School*, unpublished MA thesis, Chelsea College, London.

Nolder, R. (1985) 'Complex numbers', *Mathematics Teaching*, *110*, March, p. 34.

O'Shea, T. and Self, J. (1983) *Learning and Teaching with Computers*, Harvester, Brighton.

Papandropoulu, I. and Sinclair, H. (1974) 'What is a word? Experimental study of children's ideas about grammar', *Human Development*, *17*, pp. 241–58.

Papert, S. (1980) *Mindstorms: Children, Computers and Powerful Ideas*, Harvester, Brighton.

Pimm, D. (1981) 'Metaphor and analogy in mathematics', *for the learning of mathematics*, *1*(3), pp. 47–50.

Pimm, D. (1982) 'Why the history and philosophy of mathematics should not be rated X', *for the learning of mathematics*, *3*(1), pp. 12–5.

Pimm, D. (1983) 'Similarities and differences between mathematical and linguistic transformations', *Nottingham Linguistic Circular*, *12*(1), pp. 66–82.

Pimm, D. (1984) 'Who is we?', *Mathematics Teaching*, *107*, June, pp. 39–42.

Polya, G. (1948) *How to Solve It*, Dover, New York.

Pulgram, E. (1976) 'The typologies of writing systems', in W. Haas (ed.), *Writing Without Letters*, Manchester University Press, Manchester, pp. 1–28.

Recorde, R. (1557) *The Whetstone of Witte*, London.

Reddy, M. J. (1979) 'The conduit metaphor', in Ortony, A. (ed.), *Metaphor and Thought*, Cambridge University Press, Cambridge, pp. 284–324.

Ritchie, G. (1982) 'Computational approaches to language', *Nottingham Linguistic Circular*, *11*(1), pp. 1–23.

Roth, P. (1970) *My Life as a Man*, Holt, Rinehart & Winston, New York.

Rubin, A. (1980) 'A theoretical taxonomy of the differences between oral and written language', in R. J. Spiro (ed.), *Theoretical Issues in*

Reading Comprehension, Lawrence Erlbaum Associates, Hillsdale, N.J.

Schools Council (1977) *Mixed-Ability Teaching in Mathematics*, Evans/ Methuen Educational, London.

Searle, J. R. (1979) *Expression and Meaning*, Cambridge University Press, Cambridge.

Sellar, W. C. and Yeatman, R. J. (1960) *1066 and All That*, Penguin, Harmondsworth.

Shuard, H. and Rothery, A. (eds) (1984) *Children Reading Mathematics*, John Murray, London.

Sinclair, H. (1983) 'Young children's acquisition of language and understanding of mathematics', *Proceedings of the Fourth ICME*, Birkhäuser, Boston.

Sinclair, J. McH. and Coulthard, M. (1975) *Towards an Analysis of Discourse*, Oxford University Press, London.

Skemp, R. R. (1979) *Intelligence, Learning and Action*, John Wiley & Sons, London.

Skemp, R. R. (1986, 2nd edn) *The Psychology of Learning Mathematics*, Penguin, Harmondsworth.

Skett, T. (1985) 'What can you tell me about these?', *Mathematics Teaching*, *112*, September, p. 23.

Stoppard, T. (1967) *Rosencrantz and Guildenstern are Dead*, Faber & Faber, London.

Stubbs, M. (1980) *Language and Literacy*, Routledge & Kegan Paul, London.

Stubbs, M. (1983a) *Discourse Analysis*, Basil Blackwell, Oxford.

Stubbs, M. (1983b, 2nd edn) *Language, Schools and Classrooms*, Methuen, London.

Sussmann, H. J. and Zahler, R. S. (1978) 'Catastrophe theory as applied to the social and behavioural sciences: a critique', *Synthèse*, *37*, pp. 117–216.

Swan, M. (1981) Unpublished transcript.

Swan, M. (1982) *The Meaning and Use of Decimals*, available from Shell Centre for Mathematical Education, University of Nottingham, Nottingham.

Sykes, J. B. (ed.) (1982) *The Concise Oxford Dictionary*, Oxford University Press, Oxford.

Szabo, A. (1978) *The Beginnings of Greek Mathematics*, D. Reidel, Dordrecht, Holland.

Tahta, D. (1970) 'Idoneities', in ATM members (eds), *Mathematical Reflections*, Cambridge University Press, Cambridge.

Tahta, D. (1984) Review of Rucker, R. (1982), *Infinity and the Mind*, in *Mathematics Teaching*, *107*, June, p. 47.

Tall, D. O. (1977) 'Cognitive conflict and the learning of mathematics', *Proceedings of the first IGPME Conference*, University of Utrecht, Utrecht.

Taylor, W. (1984) 'Metaphors of educational discourse', in W. Taylor (ed.), *Metaphors of Education*, Heinemann, London.

Thom, R. (1971) 'Modern mathematics: an educational and philosophical error', *The American Scientist*, 59(6), pp. 695–6.

Thom, R. (1973) 'Modern mathematics: does it exist?', in A. G. Howson (ed.), *Developments in Mathematics Education*, Cambridge University Press, Cambridge.

Thorndike, E. L. (1922) *The Psychology of Arithmetic*, Macmillan, New York.

Truffaut, T. (1967) *Hitchcock*, (English trans.), Simon & Schuster, New York.

Ullmann, S. (1957) *The Principles of Semantics*, Basil Blackwell, London.

UNESCO (1975) *Interactions between Linguistics and Mathematical Education*, UNESCO, Copenhagen.

Vygotsky, L. S. (1962) *Thought and Language*, MIT Press, Cambridge, Mass.

Weizenbaum, J. (1984) *Computer Power and Human Reason*, Pelican, Harmondsworth.

Wheeler, D. (1983a) 'Mathematics and language', in C. Verhille (ed.), *Proceedings of the Canadian Mathematics Education Study Group*, 1983 Annual Meeting, pp. 86–9.

Wheeler, D. (1983b) 'Some problems that research in mathematics education should address', *Proceedings of the seventh PME Conference*, Shoresh, Israel.

Whitehead, A. N. (1925) *Science in the Modern World*, Macmillan, New York.

Whitehead, A. N. (1969) *Process and Reality: an Essay in Cosmology*, Free Press, New York.

Wilder, R. L. (1968) *The Evolution of Mathematical Concepts*, Wiley, London.

Wills, D. D. (1977) 'Participant deixis in English and baby talk', in C. E. Snow and C. A. Ferguson (eds), *Talking to Children*, Cambridge University Press, Cambridge, pp. 271–308.

Wing, T. (1985) 'Reading Children Reading Mathematics', *Mathematics Teaching*, 111, June, pp. 62–3.

Winner, E. (1979) 'New names for old things: the emergence of metaphoric language', *Journal of Child Language*, 6, pp. 469–91.

Wittgenstein, L. (1958) *Philosophical Investigations*, Basil Blackwell, London.

Woolf, V. (1945) *A Room of One's Own*, Penguin, Harmondsworth.

Yates, J. (1978) *Four Mathematical Classrooms*, Technical Report, available from Faculty of Mathematical Studies, University of Southampton, Southampton.

Videotapes

EM235 *Developing Mathematical Thinking* (1982).
Working Mathematically with Low Attainers (1985).
PM644 *Secondary Mathematics: Classroom Practice* (1986).
(Details of these videotapes may be obtained from the Centre for Mathematics Education, Faculty of Mathematics, The Open University, Walton Hall, Milton Keynes, Bucks MK7 6AA, England.)

Index